PERMACULTURE AND CLIMATE CHANGE ADAPTATION

Inspiring Ecological, Social, Economic and Cultural Responses for Resilience and Transformation

Thomas Henfrey
and Gil Penha-Lopes

Permanent Publications

Overhead view of a Jatropha plant
CAROLYN MONASTRA

Editors and lead authors:
Thomas Henfrey and Gil Penha-Lopes
Center for Ecology, Evolution and Environmental Change (cE3c)
CCIAM research group
Faculty of Sciences, University of Lisbon

Design and illustrations:
Ruth Wellman, TinCat Design (www.tincatdesign.co.uk)

Additional contributors:
Filipe Alves (cE3c research group, Faculty of Sciences, University of Lisbon)
Hugo Oliveira (cE3c research group, Faculty of Sciences, University of Lisbon)
Katy Fox (Centre for Ecological Learning Luxembourg)
Myriam Legault (Mesoamerican Permaculture Institute, Guatemala)
Pierre Monsoor (Moringa, Malawi)
Terrence Leahy (University of Newcastle, Australia)
Graham Brookman (Food Forest, Australia)
Naomi van der Velden (Permaculture Association, Britain)
Chris Warburton Brown (Permaculture Association, Britain)
Andre Vizinho (cE3c research group, Faculty of Sciences, University of Lisbon)
Jerome Osentowski (Central Rocky Mountains Permaculture Institute, USA)
Chris Evans (Himalayan Permaculture Centre, Nepal)
Sophie (TERRA CSA)

Thanks to:
Carolyn Monastra (Witness Tree), Starhawk, Iga Gozdowska (Climate Futures), Zipporah Lomax, May East (Gaia Education), Martin Stengel (Sieben Linden), Maddy Harland (Permanent Publications), Tim Harland (Permanent Publications), Sandra Campe (Permakultur Akademie), Gregory Landua (Terra Genesis), Monica Picavea (Transición Brasilandia), Andrew Langford (Gaia University), Albert Bates (The Farm, Tennessee), Andy Goldring (Permaculture Association), Eamon O'Hara (AEIDL), Stella Strega (Integral Permaculture Academy)

Pictures generously provided by:
Carolyn Monastra, Witness Tree (www.thewitnesstree.org)
Chikukwa Project (www.chikukwa.org)
Zipporah Lomax (http://zipporah.smugmug.com; www.dustyplayground.com)
Iga Gozdowska (http://igagozdowska.com)
Food Forest (www.foodforest.com.au)
Gaia Education (www.gaiaeducation.org)
Mesoamerican Permaculture Institute (http://imapermaculture.org)
Kusamala Institute of Agriculture & Ecology (www.kusamala.org)
African Moringa and Permaculture Project (www.ampp.org.uk)
Climate Futures (www.climatefutures.co.uk)
The Orskov Foundation (www.orskovfoundation.org)
The James Hutton Institute (www.hutton.ac.uk)
Terra Genesis (www.terra-genesis.com)
Permaculture Association Britain (www.permaculture.org.uk)

Editorial input from Gil Penha-Lopes and printing were financially supported by Fundação para a Ciência e Tecnologia (FCT Unit funding, ref. UID/BIA/00329/2013; and SFRH/BPD/65977/2009).

FCT Fundação para a Ciência e a Tecnologia
MINISTÉRIO DA EDUCAÇÃO E CIÊNCIA

The Schumacher Institute

"This project has received funding from the European Union's Seventh Framework Programme for research, technological development and demonstration under Grant Agreement No. 308337 (Project BASE). The contents of this website are the sole responsibility of BASE and can in no way be taken to reflect the views of the European Union." http://base-adaptation.eu

CAROLYN MONASTRA

Published by:
Permanent Publications
Hyden House Ltd
The Sustainability Centre
East Meon
Hampshire GU32 1HR
United Kingdom
Tel: +44 (0)1730 823 311
Fax: 01730 823 322
Email: enquiries@permaculture.co.uk
Web: www.permanentpublications.co.uk

and:

Good Works Publishing Cooperative,
c/o The Schumacher Institute, Create Centre,
Smeaton Road, Bristol BS1 6XN
www.goodworkspublishingcoop.wordpress.com

Distributed in the USA by:
Chelsea Green Publishing Company, PO Box 428,
White River Junction, VT 05001
www.chelseagreen.com

Printed in the UK by CPI Antony Rowe, Chippenham, Wiltshire

All paper from FSC certified mixed sources

The Forest Stewardship Council (FSC) is a non-profit international organisation established to promote the responsible management of the world's forests. Products carrying the FSC label are independently certified to assure consumers that they come from forests that are managed to meet the social, economic and ecological needs of present and future generations.

British Library Cataloguing-in-Publication Data
A catalogue record for this book is available from the British Library.

ISBN 978 1 85623 275 3

CONTENTS

Women carrying wood in Dowa district, Malawi. Climate Smart Agriculture for smallholder farmers in Malawi project. 2015 IGA GOZDOWSKA

FOREWORD

Climate change calls for us to think differently about adaptation. This book does precisely that – it shows how society can pursue effective climate change adaptation through permaculture, a concept that embodies the vision of a sustainable world, or a "permanent culture." Drawing attention to diverse examples, Henfrey and Penha-Lopes describe a growing global grassroots movement that is practicing adaptation to climate change through much wider and deeper transformations to sustainability. This movement is dynamic, holistic, self-organizing, and taking place outside of the mainstream sustainability discourse, which struggles hard to address climate change through an outdated paradigm.

This book goes beyond that paradigm to reveal what a more holistic, bottom-up approach has to offer. Rather than prescribing a fixed recipe for sustainability, permaculture includes a diversity of concepts, knowledges, strategies, tools, techniques and practices that are reshaping the world and providing compelling visions of what is possible. By focusing on the patterns, structures and dynamics of living systems, we see clear examples of how we can restructure, regenerate, restore, and renew in ways that actually make a difference.

The impacts and implications of these grassroots activities for climate change adaptation have been inadequately studied through formal research, yet there is no doubt that this is the type of knowledge that we need right now. This is a refreshing view of climate change adaptation – one that is exciting, inspiring, and engaging, and one that calls on us to step up to the adaptive challenge of climate change.

Karen O'Brien
University of Oslo, Norway

Working at Rancho Mastatal, a permaculture community spread out over 550 acres. Mastatal, Costa Rica, 2011 CAROLYN MONASTRA

1.1

INTRODUCTION

Often under the radar of policy processes, media coverage and public attention, efforts by self-organised groups of concerned citizens to take meaningful action on climate change in the context of their everyday lives are increasing in number. A 2013 study by AEIDL, the European Association for Information on Local Development, identified at least 2000 community-based initiatives for mitigating and/or adapting to climate change in EU countries;[1] subsequent research suggests this conservative figure was a dramatic underestimate.

Here we seek to celebrate the vision, dedication and conviction of those who take personal responsibility for making a difference

The quiet growth of this movement contrasts with the declining prominence of climate change in both public and political consciousness since the disappointment of the 2009 COP in Copenhagen, where amidst much fanfare, little of substance was delivered. The run-up to COP 21 in Paris this autumn is marked by a similar sense of urgency: at least in the rhetoric of major environmental NGOs, who are presenting it as another make-or-break moment. Alongside this, absence of political razzmatazz reveals a worrying sense of jaded resignation about the process itself. Reports have emerged of deals brokered behind the scenes to maintain the status quo and protect vested interests. Many campaign groups have consequently abandoned hope that any meaningful agreement will be reached, with some calling for a concerted show of opposition to the entire process. Meanwhile, the plight of Syrian climate migrants – by no means the first people to lose their homes and livelihoods to climate change, but the first to seek refuge in Europe in such numbers – brings the human consequences into prominent focus. In the face of political inertia, truly terrifying implications, and urgent immediate humanitarian concerns, it is small wonder if many among the general public prefer not to think too deeply or often about climate change.

Here we seek to celebrate the vision, dedication and conviction of those who take personal responsibility for making a difference. Common to many community-based responses to climate change – not just in Europe, but worldwide – is use of permaculture as both guiding philosophy and tool for practical action. Permaculture is gaining prominence as a term – a Google search for the word in October 2015 turned up nearly 8 million hits – but is far less widely understood. It is widely perceived as some sort of ecological agriculture: while this is an important area of application, its full scope is far broader. This book seeks to convey some of this breadth, and permaculture's consequent importance as one among the many different tools and approaches necessary if the global response to climate change is not only to avoid disaster, but perhaps herald new ways of living within planetary limits as a responsible global society.

Bangladeshi women carrying out a process of measuring the raised vegetable-growing bed by using their arm's lengths. BASD

This brief overview of how people worldwide are using permaculture in this way has five sections:

Section two, immediately following this introduction, introduces key guiding concepts: climate change and environmental stewardship, permaculture as a holistic methodology, and the coordinated research efforts necessary to fulfil this.

Section three outlines key orienting perspectives: of indigenous peoples and others living close to nature and hence to the immediate consequences of climate change, and of global science and the political community it informs.

Section four describes 17 key strategies we have identified in permaculture-based responses to climate change.

Section five considers ways forward, examining the scope for constructive engagement of permaculture and policy and describing how the permaculture movement itself is organising to work more effectively for positive change.

We hope that writing this book will contribute in some way to these efforts, and support both policy responses and self-organised efforts. We dedicate our work to all those working to transform climate change from a dangerous threat into an inviting opportunity to create a better world.

2. BACKGROUND

Climate change is more than simply a technical issue. It is an invitation to re-assess humanity's place in the world, and to transform global society in ways that allow our continued survival. Permaculture originated in just such a re-assessment, and has become a significant impetus for such a transformation.

This section examines permaculture's relationship to current scientific thought on the implications of climate change, which is converging on similar conclusions, and considers how closer engagement of permaculture and climate science can help support meaningful action.

One of the valleys of the Chikuwka clan, Zimbabwe, in 2014. In the foreground we can see houses nestled within their orchards and cropping fields. Further off we can see how woodlots have been established on the ridge tops and fields are supported by contour bunds. In the distance is the forestry land with its firebreak. Chikukwa Project. TERRENCE LEAHY

2.1

CLIMATE CHANGE: FROM ADAPTATION TO TRANSFORMATION

As understanding of the significance of climate change deepens, the view that responses will involve a transformation in human relationships with nature becomes increasingly widespread

In the present day, climate change is no longer a hypothetical future possibility, but an inescapable fact of everyday life. As climatologists become more certain about human effects on global atmospheric composition and their consequences, extreme weather events become ever more common and slower trends such as sea level rises and changes in seasonal weather patterns continue. The most recent summary report of the Intergovernmental Panel on Climate Change reaches some stark conclusions.[1] It predicts, with high levels of certainty, continued rises in global mean surface temperatures if greenhouse gas emissions are not abated, and alongside this greater and more frequent extremes of heat, global increases in precipitation, and continued loss of Arctic sea ice. It also suggests that continued changes in many aspects of global climate systems are likely even if temperatures stabilise, and raises the possibility of abrupt shifts in some of these.

This uncomfortable prospect has led to global calls to mitigate the causes of climate change, through deep and rapid reductions in greenhouse gas emissions along with mechanisms to remove and sequester carbon dioxide already in the atmosphere.[2] At the same time, governments of the wealthiest nations continue to provide subsidies to oil, gas and coal companies for fossil fuel production, estimated to total around USD88 billion during the year 2014.[3] Despite the announcement by G7 nations of an agreement to phase out fossil fuels by 2100, the prospects for timely mitigation remain uncertain at best.

With ongoing significant climate change inevitable, recent years have seen a shift in emphasis from mitigation to adaptation: reducing the vulnerability of ecological and human systems to climate change impacts, and taking advantage of the opportunities these impacts present.[4] The EU has produced its own climate change adaptation strategy,[5] and a web platform, Climate-Adapt, to support this.[6] At the local level – where most adaptation measures will need to be implemented – European-wide networks of municipal authorities such as the Covenant of Mayors, Mayors Adapt and ICLEI have taken various types of action, including the Annual Global Forum on Urban

This shows a typical scene from the land of the Chikukwa Clan in the early nineties when the project started. CHIKUKWA PROJECT/TERRENCE LEAHY

Resilience and Adaptation that since 2012 has brought together decision makers, municipal technicians, academics and NGOs to share, learn and build partnerships for designing and implementing effective local solutions. Calls are increasing for more radical approaches that go beyond specific isolated mitigation and adaptation measures and embrace the need for systemic change.[7] This change includes reconfiguration of material systems, and also goes beyond it to psychological and cultural transformations: in our perceptions and shared understandings of humanity's relationship to the natural world.[8] As such, climate change represents more than just a threat to the continuation of society as currently experienced by most people in the world. It is an invitation to accept the necessity for transformation, and embrace it as an opportunity to reimagine and reshape our world for the better, in line with the responsibility that comes with technological power that is – quite literally – earth-shattering.

This book considers the possibilities and realised achievements of permaculture, as an existing framework for analysis and action that can help achieve such a transformation. Going beyond conventional notions of sustainability, permaculture accepts change as an inevitable feature of existence, and provides conceptual and practical tools to negotiate and undertake change creatively and proactively.[9]

As later sections show, in its nearly four decades of existence permaculture has grown into a flourishing and vibrant worldwide movement operating largely at the social, political and economic margins. Its strategies for integrating and working with change are wide-ranging. They cover social, cultural and psychological as well as material dimensions, and have been applied in all types of settings, from remote forests and deserts to the largest cities, all over the world. A key methodology in environmental movements such as ecovillages, bioregionalism, and Transition Towns, it is increasingly evident that it is a vital tool in addressing the collective challenges humanity currently faces.

2.2

PERMACULTURE AND CLIMATE CHANGE ADAPTATION

Permaculture provides an ethically grounded framework for designing and implementing holistic responses to climate change with existing and potential applications worldwide

Permaculture is a design system for sustainable human habitats, taking an ecological perspective and rooted in an explicit ethical framework. It was originally conceived in the 1970s by Australian field ecologists David Holmgren and Bill Mollison as a contraction of the term 'permanent agriculture'.[1] It later expanded its scope to encompass the full range of factors affecting the ecology of human settlement, economy and culture, and is now more commonly considered shorthand for 'permanent culture'.

The basic philosophy is one of working with rather than against nature, designing human habitats and organisations in ways that deliberately seek to emulate features that contribute to resilience, sustainability and productivity in natural systems. It thus has much in common with other approaches that take inspiration from nature in the conscious design of human systems, including biomimicry,[2] ecological engineering,[3] and adaptive management (which in turn has many features in common with indigenous and traditional environmental management systems).[4]

Permaculture locates itself at the intersection of three mutually interdependent ethics: Earth Care, People Care and Fair Shares. Its take on climate change is hence both environmentalist and humanist, proposing responses that take into account responsibilities for the planet's biota as a whole while also paying attention to social justice and equity. This places it in opposition to social, political and economic systems that rely on ongoing ecological degradation, leave basic human rights unfulfilled for large numbers of people, and systematically increase inequalities of wealth and power.[5] It seeks not only to arrest these trends, but to reverse them, actively finding and enacting possibilities for social and ecological restoration.

Permaculture's ecological orientation, its ethical basis and its transformative implications all anticipate the practical and moral conclusions of recent debates sparked by climate science. Key among these is the suggestion that climate change marks the onset of the Anthropocene, a new geological epoch in which the impacts of human activities on the biosphere are as significant as those of natural processes.[6]

Scale model of the Food Forest, a permaculture farm and learning centre on the Adelaide Plain in Australia. GRAHAM BROOKMAN

Climate change and other major negative impacts such as loss of biodiversity, soil and fresh water threaten to disrupt global environmental conditions to the extent they are no longer conducive to supporting human civilisation.[7] Earth systems scientists from many different disciplines now argue for a shift in the premise of global governance: from treating the earth's natural wealth as resources to be exploited, to an attitude of conscious stewardship of planetary systems.[8]

Working precedents for such an ethos of environmental stewardship can be found in the many ways indigenous and traditional societies have ecologically enriched the lands they inhabit. Distributions of cultural and biological diversity are strongly correlated across most of the globe.[9] Many highly ecologically rich areas previously thought of as pure wilderness are now known to be cultural products whose biotic features reflect centuries of previous human management.[10] Permaculture explicitly cites such societies as models for simultaneously enhancing biotic and productive qualities (Chapter 3.1). Some indigenous groups have adopted permaculture as a strategy for safeguarding and advancing traditional knowledge and management systems (Chapter 4.11). Closely associated with movements for bioregionalism and resource

Some indigenous groups have adopted permaculture as a strategy for safeguarding and advancing traditional knowledge and management systems

Permaculture course at the CELL/TERRA community farm, Luxembourg. PEACE ADVOCATE PHOTOGRAPHY

localisation (Chapter 4.8), permaculture has become a key tool for efforts to reconnect human activities with their geographical resource base. It does this, in part, by learning from peoples who never lost that connection.

Living in sensitive habitats, directly dependent on the local ecology, and often having historical experience of changes in local and regional weather patterns resulting from land use changes, many indigenous societies have needed to be pioneers in adapting to climate change. In parallel fashion, permaculturists often work in places damaged in ways that prefigure the effects of climate change: where deforestation, land degradation, and socio-economic disruption have already severely undermined local ecologies and livelihoods. While indigenous experiences of climate change may predate global responses by centuries, those of permaculturalists may be decades ahead. Both hold important lessons for wider action.

The purpose of this book is to illustrate some of the ways in which permaculture is being used to support adaptation to climate change worldwide. It describes strategies developed over four decades of hands-on experience of conceiving and implementing practical responses to environmental and social change, and some of their current applications relevant to climate change adaptation. These strategies and associated techniques are diverse, and social as well as material in nature. Those described here represent only a fraction of those actually employed in permaculture, a movement of thousands of trained practitioners in virtually every country in the world, whose approaches are as varied as their personalities, backgrounds, and the circumstances in which they work.

Sustainable Rice Intensification planting in Nepal. NAOMI VAN DER VELDEN.

2.3

CULTIVATING A GLOBAL CLIMATE CHANGE ACTION RESEARCH COMMUNITY

Linking permaculture's action learning community more closely with formal science is vital to realising its full potential to contribute to global action on climate change

While permaculture is extensively practiced worldwide, formal documentation of the sort that could inform policy is limited. Data on the impacts of permaculture are scarce. This restricts both its capacity for critical self-evaluation and its ability to provide robust evidence to policy-makers and other interested stakeholders. Changing this situation is crucial if permaculture is ever to realise its full potential to contribute to responses to climate change.

Many techniques commonly used in permaculture are relatively well documented. For example, there is an extensive literature on agroecology.[1] The sustainable rice intensification approaches employed by the Himalayan Permaculture Centre are grounded in extensive scientific research.[2] However, very little robust data exists on permaculture itself, either as a design methodology or community of practice.

Resource limitations severely constrain practitioners' capacity for self-documentation and self-evaluation. The projects who contributed information to this book consistently reported a need for more and better research on their work. Needing to prioritise practical outcomes when allocating scarce labour and funds, few are in a position to deliver this themselves.

Formal academic studies have increased in recent years, but are still small in number relative to the size of the movement.[3] Often a clash of cultures restricts collaboration between permaculturists and professional researchers whose horizons are limited by concerns with academic prestige, intellectual fashions, and the requirements of conservative funding regimes. This is an instance of the permaculture principle, "The solution is in the problem": it reveals ways in which permaculture can help research become more relevant and effective.[4]

Permaculture is inherently experimental, based on designing and testing out practical solutions to existing problems. A key feature is action learning: reflecting and critically assessing the outcome of any intervention, and taking this into account in subsequent action. Through teaching and writing, this learning spreads to the wider permaculture community. In this way,

Transition Research Network Workshop, Plymouth, UK, Feb 29th 2012. The board in the background shows a concept Permaculture Design for the network. TRANSITION RESEARCH NETWORK

permaculture practitioners taking practical action for climate change adaptation worldwide form an informal global action learning community.

As this community grows in number and capacity, the permaculture movement worldwide is starting to take responsibility for its own documentation and evaluation.[5] Academic researchers in applied fields and wanting their work to contribute to meaningful social change have highlighted the affinity with permaculture.[6] The frequency and sophistication of collaborations between researchers and permaculturists are increasing, and more people combine formal academic qualifications with training in permaculture.[7]

Globally, the UK Permaculture Association has taken a leading role. In the view of Research Coordinator Dr. Chris Warburton Brown, permaculture practitioners are not research subjects, but active researchers in their own right, whose knowledge and skills are as important as those of professional researchers. Influenced by approaches such as citizen science and participatory action research, the Association involves permaculturists as co-researchers in participatory field trials on polycultures, forest gardens, changes in soil and biodiversity, and cultivation of wheat and soya.[8] It has produced a Research Handbook for practitioners, providing a clear non-technical guide to conducting meaningful research projects.[9] This work reaches wider audiences through vehicles such as the Permaculture Research Digest,[10] which receives over 2,500 page views every month, and the International Permaculture Conference in London in 2015,[11] the most ambitious permaculture research event ever held.

This work takes place under severe financial constraints. Conservative and inward-looking attitudes on the part of academic funders mean it has so far been financed exclusively from a small number of philanthropic sources.

The new Permaculture International Research Network (PIRN) will bring together permaculture researchers worldwide into a single online network

Resource limitations oblige Chris to take a creative approach to his work: a form of social permaculture (Chapter 4.14) that maximises the outputs from very limited financial inputs. Strategies include reliance on student volunteers, a focus on online rather than physical outputs, and collaborating with the Association's membership network to minimise the cost of collecting field data.

The Association's latest project responds to the need to upscale permaculture research. The new Permaculture International Research Network (PIRN) will bring together permaculture researchers worldwide – both academics and practitioners – into a single online network for mutual support, learning, sharing and collaborative research projects.[12] PIRN was conceived on the basis of four online surveys of the global permaculture research community, which assembled an initial membership of nearly 400 researchers in over 50 countries and fed into creation of a development plan prior to the Network's formal launch at the International Permaculture Conference in London in September 2015. Although a crucial dimension of permaculture's prospects of fulfilling its potential contributions to addressing climate change, PIRN's continued existence is precarious and will depend on its ability to mobilise core funding and leverage wider changes in how research is funded and undertaken.

Some formal research projects are already initiating such changes, integrating Permaculture ethics and principles into their scientific methodologies, often employing participatory action research.[13] The EU FP7-funded research project 'Bottom-up Climate Adaptation Strategies Towards a Sustainable Europe' (BASE)[14] employs permaculture-trained co-researchers on design and implementation of the Portuguese case study, significantly improving its impact, depth and scientific quality. Use of Systematisation of Experiences[15] methodologies supported self-reflection around the 'Earth Care' ethic by both project staff and projects collaborating as case studies. Co-creation of outputs by researchers and practitioners became a further vehicle for collaborative learning. Personal development coaching, monthly team awaydays, a collaborative approach to case study research and involvement of members of study initiatives in the national advisory board, all contribute to bringing the 'People Care' ethic to the heart of the project's research culture. Holding funded scientific events at Permaculture projects brought financial support to local initiatives, [16] an expression of the 'Fair Shares' ethic. In a similar fashion, the Transition Research Network uses permaculture-based methods to enable researchers and practitioners in the Transition movement (Chapters 4.8 and 4.9) to collaborate more effectively.[17]

Both BASE and the Transition Research Network have taken active roles in establishment of ECOLISE, a network of European-based community-led sustainability initiatives whose core members include national permaculture associations and closely related networks in Transition and ecovillages.[18] Arising in part from a failed effort to secure FP7 funding for a highly collaborative project involving permaculture-based organisations as co-researchers, ECOLISE has identified Knowledge and Learning among its four main pillars of work.

This recognises the need both for self-reflection and learning, and more effective communication with policy audiences and other external stakeholders. The final round of EU FP7 projects starting in 2013 includes two concerned with mapping, characterising and evaluating community-based sustainability initiatives, including permaculture projects: TESS (Towards European Societal Sustainability)[19] and ARTS (Accelerating and Rescaling Transitions to Sustainability).[20] ECOLISE's Knowledge and Learning Working Group actively collaborates with both, to maximise the value and impact of this vital work and to open pathways to more engaged, practice-based research in the future.

Caring for rice seedlings cultivated using sustainable rice intensification techniques. Himalayan Permaculture Centre.

3. PERSPECTIVES, LOCAL AND GLOBAL

While climate change is a global phenomenon, its effects are most meaningfully felt at local and bioregional scales. Awareness of climate change as a global issue is relatively recent, having become widespread only in recent decades. However, many indigenous and traditional peoples have experience of climatic changes over longer time scales, providing perspectives that complement those from earth systems science and culturally embedded capacities to respond.

This section looks at local and global perspectives and considers permaculture's role in linking the two. Indigenous lands, cultures, lifestyles and knowledge are under ongoing threat worldwide, seriously weakening the global human capacity to adapt to climate change. Permaculture counters this, both directly where it provides a tool for strengthening and safeguarding indigenous cultures (Chapter 4.11), and indirectly where it support new social and cultural innovations better suited to changing circumstances (Chapters 4.10, 4.12 and 4.14).

Chief Khengsamut gives presentations to all tourists who visit to impress upon them how much sea level rise has affected their village: residents have had to move their school and homes inland four to five times in the past few decades as waters continue to rise and erode their shoreline. Thailand, 2012.
CAROLYN MONASTRA

3.1

INDIGENOUS PEOPLES, CLIMATE CHANGE AND PERMACULTURE

Permaculture seeks to emulate many features of indigenous and traditional environmental management systems, which often reflect long histories of adaptation to environmental change

There are many parallels and direct linkages between permaculture and indigenous livelihood strategies. Permaculture in part originated from observations of how indigenous peoples in Tasmania and elsewhere relate to their environments, and continues to learn from indigenous knowledge. Some indigenous groups successfully employ permaculture as a tool in adapting traditional environmental management strategies to changing circumstances (Chapter 4.11). Understanding indigenous experiences of and responses to climate change can inform both the practice of permaculture and policy-level efforts to support it.

There is no simple or uncontroversial way to define 'indigenous'. For present purposes, we can take it to refer to people who depend for basic subsistence upon goods and services provided by their immediate biotic environment.[1] The practicalities of survival in such circumstances in certain ways prefigure, and so can inform, emerging global challenges around sustainability and climate change adaptation.[2]

Conservation is a western concept, without known parallel in any documented non-European language.[3] However, people who lack the option either to externalise environmental damage or to migrate away from its effects have historically needed to maintain and enhance their resource base for the long term. Strategies for achieving this are as varied as the environments people inhabit.[4] The concept of Biocultural Diversity describes the strong correlations between biological and cultural diversity that have been documented over much of the globe.[5]

A common feature of diverse indigenous resource management systems is their flexibility and adaptability in the face of change. Variability, uncertainty and unpredictability are inherent in any complex ecological formation; even more so in social-ecological systems in which people play a major role. Indigenous resource management strategies typically accommodate and respond to change rather than limit or control it. They achieve this through sophisticated mechanisms for monitoring and responding to changes in ecological conditions and environmental feedbacks. These include ways to cope with natural variability in weather, including rare extremes of temperature, drought, flood, and wind, even if these arise only at intervals of several decades or generations. Existing mechanisms for navigating inherent environmental unpredictably and coping with rare severe weather events can be a basis for long-term adaptation to permanent changes in weather conditions.[6]

Permanent changes in local or regional climates are within the historical experience of many contemporary indigenous peoples

Permanent changes in local or regional climates are within the historical experience of many contemporary indigenous peoples, particularly those exposed to the impacts of colonial or state projects with large scale impacts on vegetation cover.[7] Removal of forests, for example, can significantly alter precipitation patterns and seasonal flows of watercourses. Large scale abstraction of water from river systems for irrigation, to supply urban centres or by hydroelectric dams affects the availability of water downstream. Many indigenous societies have embedded collective learning from such experiences in social or cultural capacities for navigating change potentially invaluable in climate change adaptation.

Many present-day indigenous populations are among those most directly affected by climate change, and most aware of it.[8] Fishing and hunting among Arctic peoples are closely attuned to seasonal cycles of changing ice cover and movements of animals.[9] Many subsistence farmers live in marginal habitats sensitive to changes in annual patterns of rainfall or temperature, or potentially vulnerable to outbreaks of fire and of crop pests and diseases. While many indigenous populations are therefore particularly susceptible to the impacts of climate change, they are also in many ways uniquely equipped to adapt.[10] Adaptation efforts elsewhere have much to learn from this, and the actions of many permaculturalists working at grassroots level already do so.

To overstate the parallels between indigenous peoples and permaculturalists would be to simplify in ways that misrepresent both.[11] Despite important differences in both historical experience and contemporary circumstances, there is significant common ground and scope for mutual learning, which in turn can inform wider efforts to address climate change.[12] The worldwide permaculture movement can be viewed as a set of deliberate experiments in the creation of biocultural diversity. It includes among its key strategies agroecology (Chapter 4.5), whose scientific basis lies in documentation of indigenous horticultural systems characterised by high diversity of techniques and yields and close integration with ongoing ecological processes.[13] Permaculture also has close associations with movements for bioregionalism

CAROLYN MONASTRA

and localisation (Chapter 4.8), which among other things seek to strengthen feedbacks between production and consumption and their ecological consequences:[14] thus emulating the integration with local ecological systems characteristic of indigenous production systems. Within Europe, similar parallels are evident in comparison of permaculture with Romanian peasant agriculture during its post-socialist adaptation to rapidly changing political, economic, social and cultural conditions,[15] and in integration of traditional and modern agriculture practices on Latvian eco-farms.[16]

Increasing numbers of practical initiatives directly link permaculture with traditional agriculture and environmental management practices (Chapter 4.11). Many permaculture projects also seek to create new forms of commons-based governance (Chapter 4.10), and so emulate a key feature that allows close and flexible attunement to local conditions in indigenous resource management.[17] Key examples referred to in this book include the Chikukwa Project in Zimbabwe, where subsistence farmers use permaculture as an integrating framework for innovation in traditional farming practices, increasing yields and food security, restoration of natural vegetation for landscape stabilisation and protection of watersheds, conflict resolution, and female empowerment.[18] The Mesoamerican Permaculture Institute was founded by a group of Kakchiquel Maya in response to increasingly severe social, cultural and environmental problems in Guatemala, and works with local farmers within a framework that combines permaculture with traditional practices and worldviews.[19]

As temperatures rise and drought becomes more widespread in Northeast China, areas of former pastureland are turning into desert. Inner Mongolia, China, 2012.
CAROLYN MONASTRA

CAROLYN MONASTRA

3.2

GLOBAL PERSPECTIVES: BIOMES, PLANETARY BOUNDARIES AND SUSTAINABLE DEVELOPMENTAL GOALS

Global climate science complements local perspectives on climate change action, with key concepts such as biomes and planetary boundaries informing international initiatives like the Sustainable Development Goals

Complementing the bounded local visions of indigenous peoples and others living in intimate contact with their local ecology, Earth Systems Science and related fields provide the broader view necessary to understand climate change as a global issue affecting all humanity. Key relevant concepts explored in this chapter include Biomes and Planetary Boundaries – which help to understand human impacts on the environment at global scale – and the Sustainable Development Goals (SDGs), a new framework for global action introduced by the UN. Permaculture can help with both the implementation of SDGs in their current form, and with future steps for their improvement.

Biomes are broad habitat and vegetation types, roughly at continental scale, that provide a basic high-level categorisation of global ecological diversity. They can be delineated in many different ways; the most widely used of these, developed by the World Wide Fund for Nature, identifies 14 terrestrial biomes worldwide. Based on their original or potential extent (according to original ecological and climatic conditions, and not taking human-induced land cover changes into account), these biomes vary in total area from nearly 35 million square kilometres (deserts and other arid zones) to 350,000 square kilometres (mangroves). They are widely dispersed geographically – aside from Oceania, which is mostly made up of small islands dominated by tropical grasslands and forest, most of the world's major biogeographic regions include from nine to 11 of the 14 biomes. This makes the concept very useful for examining human ecological impacts on a global scale.

The Millennium Ecosystem Assessment found that, by 1990, human activity had reduced

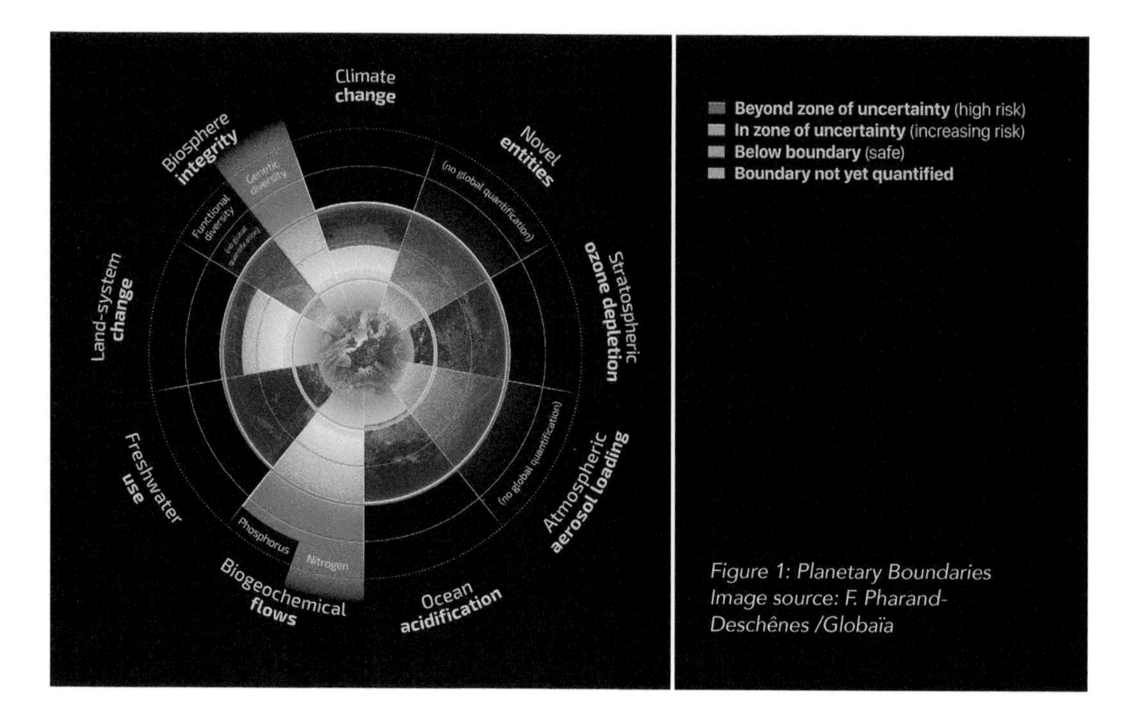

Figure 1: Planetary Boundaries Image source: F. Pharand-Deschênes /Globaïa

the area of two of the major terrestrial biomes (Mediterranean Forests and Temperate Grasslands) by more than two thirds and another four by around half, mainly through conversion to agriculture.[1] Deserts, montane grasslands, boreal forest and tundra have experienced far less conversion: less useful for agriculture, they also have a higher proportion of their area under legal protection than other biomes. While cultivated lands provide many provisioning services such as grains, fruits, and meat, conversion to conventional agriculture typically leads to reductions in native biodiversity and loss of ecosystem services. Permaculture directly challenges this conflict, seeking to design habitats that are simultaneously abundant in terms of biodiversity, ecological richness, and production for direct human use.

An important conceptual framework for understanding global human ecology emerged from the Planetary Boundaries studies led by the Stockholm Resilience Centre (Figure 1).[2]

Conclusions of recent research confirm and broaden concerns raised by the IPCC. A 2015 report concludes that humanity has crossed four of the nine planetary boundaries: climate change, loss of biosphere integrity, land-system change and altered biogeochemical cycles (phosphorus and nitrogen). Ocean acidification is approaching its boundary while stratospheric ozone depletion, freshwater use, atmospheric aerosol load and release of novel materials are still within calculated safe limits.[3] The authors of the new report are keen to present it in a positive light: as supporting the necessary global transformation in human-environment relations. Their argument is that knowing the Earth's 'safe operating space' provides vital context for understanding what that transformation will involve, and considering how it might come about.[4]

In relation to action, UN processes around the Millennium Development Goals (MDG) and Sustainable Development Goals (SDG)

give cause for guarded optimism. The MDGs were launched by the UN in the year 2000 and signed by a final total of 193 countries and 23 International organisations. Until 2015 they defined the basic development goals promoted and funded through the UN and other international processes.[5] The eight goals were:

1. To eradicate extreme poverty and hunger
2. To achieve universal primary education
3. To promote gender equality
4. To reduce child mortality
5. To improve maternal health
6. To combat HIV/AIDS, malaria, and other diseases
7. To ensure environmental sustainability
8. To develop a global partnership for development

Operationalisation of the MDGs involved 21 specific targets and more than 60 indicators. By 2015, despite great success in some terms (the majority of overall targets had been achieved – or closely missed), many problems had become clear. These include a focus on symptoms rather than underlying causes, lack of attention to key areas such as equity and environmental sustainability, and uneven progress across and within countries (masked by the use of global indicators).[6] Highly centralised processes threatened the legitimacy of the goals, which were developed with a marked lack of transparency and emphasised external initiative rather than bottom-up, participatory methods for delivery.

As this book is being written, final agreements are being made on the Sustainable Developmental Goals (SDGs), the successor to the MDGs, due to be ratified in September 2015 and come into force from January 2016. The SDGs address some of the key criticisms of the MDGs: they are intended to focus on causes as well as symptoms, and were developed through largely transparent processes, including the largest consultation in UN history. They comprise 17 goals, which translate to a total of 169 targets.

The goals are:

- End poverty in all its forms everywhere
- End hunger, achieve food security and improved nutrition, and promote sustainable agriculture
- Ensure healthy lives and promote wellbeing for all at all ages
- Ensure inclusive and equitable quality education and promote lifelong learning opportunities for all
- Achieve gender equality and empower all women and girls
- Ensure universal availability and sustainable management of water and sanitation
- Ensure universal access to affordable, reliable, sustainable and modern energy
- Promote sustained, inclusive and sustainable economic growth, full and productive employment, and decent work for all
- Build resilient infrastructure, promote inclusive and sustainable industrialisation, and foster innovation
- Reduce inequality within and among countries
- Make cities and human settlements inclusive, safe, resilient and sustainable
- Ensure sustainable consumption and production patterns
- Take urgent action to combat climate change and its impacts
- Conserve and sustainably use the oceans, seas and marine resources for sustainable development
- Protect, restore and promote sustainable use of terrestrial ecosystems, sustainably manage forests, combat desertification and halt and reverse land degradation, halt biodiversity loss
- Promote peaceful and inclusive societies for sustainable development, provide access to justice for all and build effective, accountable and inclusive institutions at all levels
- Strengthen the means of implementation and revitalise the global partnership for sustainable development

Observers have raised concerns that the transparency and inclusivity of the process was undermined by last-minute brokering and brinkmanship behind closed doors, leading to significant amendments in the wording of some key points.[7] More fundamentally, the goals are framed within an uncritical assumption that continued economic growth is both desirable and necessary.[8] This contradicts the widespread acceptance in sustainability economics that economic growth is beneficial only in specific circumstances and continued growth in the global economy is for the most part doing more harm than good.[9]

While far from perfect, the SDGs are in most respects a welcome development, and permaculture has a key role to play in their implementation and improvement. A synthesis report from UN secretary general Ban Ki-moon clusters them into six 'essential elements': dignity, prosperity, justice, partnership, planet, and people. These elements could be even more simply restated as the permaculture ethics of People Care, Fair Shares, and Earth Care. With established principles and methodologies for acting on these ethics, and numerous successful applications worldwide, permaculture offers the promise to deliver on the best intentions of the SDGs as part of a collective response to climate change that is regenerative of natural systems and transformative of human systems, emulating on a global scale the ways in which indigenous peoples have simultaneously enhanced environmental quality in ecological terms and in relation to human needs.

4. STRATEGIES

Permaculture has no prescribed set of practical techniques. Applied in settings as diverse as the Mongolian Steppes and inner city Los Angeles, it draws on a wide range of tools and approaches. Each instance of application uniquely reflects its own particular context, including the needs, interests and skills of the people involved. However, common patterns can be seen at a more general level, with identifiable strategies repeatedly employed where similar issues arise in different places and projects.

This section describes some of the main strategies documented and reported from cases where permaculture is used to support adaptation to climate change. Some are direct responses to immediate physical impacts. Others operate at a deeper level, seeking to cultivate the ecological and/or social conditions necessary for effective adaptation.

The strategies, techniques and examples in the chapters that follow are just a tiny subset of those that exist. They give an indication of what has been achieved through largely marginal action, and a tantalising glimpse of what could be achieved with suitable policy measures and delivery support.

4.1

WATER REGULATION AND MANAGEMENT

Restoring natural capacity to retain water in vegetation and soils and creating new structures to hold or divert rainfall helps farmers adapt to changing precipitation patterns

Changing climatic conditions mean a need to plan for altered and often unpredictable future precipitation patterns. In some places rainfall will decrease; elsewhere drought will be more common and/or pronounced; other areas will experience erratic rainfall, storms or other extreme weather events.[1] Applying the principle of 'capture and store energy', permaculture practitioners have designed strategies to regulate the movement of water across a site or landscape, either through controlled mechanical processes or within natural flows through soil and vegetation.[2]

A straightforward water retention technology used in many permaculture projects is swales, or earth bunds: elongated mounds dug along contours or in flat areas to intercept surface water runoff and allow its absorption in to the soil and groundwater.[3] In Guatemala, the Mesoamerican Permaculture Institute has identified swales as an element of traditional Maya water management practices, now part of a wider strategy to promote and strengthen **indigenous knowledge**.

Another simple technique is mulching: covering the soil surface with a layer of organic matter that is allowed to rot down. This reduces surface evaporation and enhances the physical capture of rainwater and surface water runoff, especially when combined with swales. It also stabilises soil surface temperatures, adds organic matter to the soil and improves habitat quality for soil organisms, and so combines water retention with **soil improvement**.[4]

The Chikukwa Project in Zimbabwe combines several techniques in an integrated strategy for water management that has reversed rapid degradation in water supplies. Creating swales and replanting natural vegetation on slopes above the fields has improved water retention in soils and reduced erosion. Areas around springs have been reforested and fenced against grazing livestock, improving availability and quality of water. Villagers have constructed simple systems to harvest and store rainwater and use household greywater to irrigate cultivated trees and vegetables. Villagers nowadays resolve disputes arising over access to water and other resources using sophisticated **conflict transformation** techniques.[5]

Widely adopted in broadscale permaculture, the Keyline Plan technique combines water retention and **soil protection**. A complex of off-contour ridges, channels and ditches directs water from potentially flood-prone sites to where it is most needed. A common addition is large-scale storage of surplus water that would otherwise run off the land, for later use by rapid, gravity powered flood irrigation.[6]

The Jordan Valley Permaculture Project has rehabilitated a four hectare desert site affected by declining rainfall, increased groundwater salinity, and erosion resulting from intermittent heavy run-off from an adjacent road. Building 1.5 kilometres of swales across the site, planting these with nitrogen-fixing and fruit-bearing trees, and extensive mulching of surrounding soils allowed establishment of initial tree cover. This was later interplanted with vegetable crops, along with grasses that improve the soil and provide forage for livestock. Crop yields match those in adjacent areas that use conventional high-input agricultural methods, with far higher efficiency of water use and steady declines in soil salinity and heavy metal content.[7]

Tamera Ecovillage in the seasonally arid Mediterranean climate of South Portugal is experimenting with landscape-scale retention of rainwater using large artificial lakes. Water slowly infiltrates the surrounding soil, replenishing aquifers. Lakeside vegetation also absorbs water, maximising evapotranspiration and condensation and increasing local humidity.[8] Tamera's water landscape project was inspired by Krameterhof in the Austrian Alps, where a network of pools and water channels supports aquaculture, irrigation, and consumption by livestock, as well as the creation of **microclimates**.[9]

Baba Matsekete shows the swales on contour in his cropping fields. Along the bund, vetiver grass holds the bund in place. The water sinks into the fields and also irrigates the orchard of the house across the slope. 2010 CHIKUKWA PROJECT/TERRENCE LEAHY

4.2

SOIL PROTECTION AND RESTORATION

Enhancing soil health addresses both symptoms and causes of climate change, improving adaptive capacity in several different dimensions while actively removing carbon dioxide from the atmosphere

Land degradation is a major cause of climate change, and degraded soils are much more vulnerable to its impacts. Declining, more erratic and/or extreme rainfall and increased rates of surface evaporation reduce available soil moisture, limiting plant and animal growth. Higher temperatures increase rates of mineralisation, reducing the soil's capacity to sequester carbon and retain water. Other potential impacts include declines in organic matter and soil biodiversity, and increased rates of compaction, erosion, and landslides.[1] Permaculture employs a range of techniques to reverse these spirals of decline: to increase the capacity of soils to store water, support a healthy underground biota, supply nutrients to plants, sequester carbon, and reduce greenhouse gas emissions, supporting both adaptation and mitigation. These techniques have in common that they mimic or enhance natural processes that help build and maintain healthy soils, using physical and biological methods. They are often combined with **revegetation** and creation of **agroecology** systems.

A common feature of forest gardens and other **agroecology** initiatives, no-till farming encompasses four broad, intertwined management practices beneficial for soil quality and adaptation of agriculture to climate change: minimal soil disturbance, maintenance of permanent plant cover, direct sowing, and sound crop rotation. Ploughing breaks up and inverts the soil: leaving it vulnerable to water and wind erosion, harming soil organisms, reducing productivity and increasing agricultural runoff; it also directly releases greenhouse gases. No-till methods cause negligible soil disturbance, opening only a narrow hole of the minimum width and depth necessary for planting. Residues from previous crops largely remain at the soil surface where they act as mulch.[2]

Another physical method is use of biochar, a charcoal-like material produced by burning fibrous plant matter at relatively low temperatures in an oxygen-scarce environment such as a kiln.[3] When buried, biochar increases the soil's capacity to retain water and nutrients and raises levels of microbial activity.[4] Burning converts the carbon in biochar into a biologically inert form that can persist in the soil for thousands of years, meaning it also has great potential for carbon sequestration and climate change mitigation.[5]

Linking physical and biological interventions, *Hugelkultur* is a traditional eastern European planting method that many permaculturalists have adopted as a soil improvement technique.

It involves covering woody debris with soil to create elevated plant beds that mimic nutrient cycling processes in natural woodlands. The decomposing wood in a hugelbed acts like a sponge, soaking up rainfall and gradually releasing it into the soil along with organic matter and nutrients, promoting fungal mycelium development and improving the general quality of the soil as a habitat for microorganisms. This technique has been extensively used at Krameterhof Permaculture Farm in the Austrian Alps, enriching the soil and creating microclimates that extend the variety and growing seasons of cultivated plants.[6]

Regenerative Agriculture, a form of broadacre permaculture developed on the US prairies and now also applied on grasslands elsewhere, employs mob grazing (or multi-paddock grazing) for soil improvement. This seeks to mimic the ecology of highly mobile herds of wild grazers by putting cattle to graze at high densities in very small areas for very short periods of time, typically a few days.[7] Grasses are eaten only to one third of their height and can regenerate rapidly once the animals move to another paddock. Partial die-back of the roots of grazed plants adds organic matter to the soil and initiates a positive feedback loop that supports soil regeneration: the soil improves, grasses grow taller and their roots deeper, and root decomposition following grazing takes place deeper in the soil. Livestock fertilise the soil with their urine and manure and partially break its surface through trampling, so nutrients are absorbed into the soil rather than being carried away by surface runoff of rainwater.[8]

Formal research has shown beneficial effects on pasture quality, soil microorganisms and water retention compared to continuous grazing.[9] This confirms anecdotal reports from ranchers practicing mob grazing across North America,[10] and practitioners who have used it to revitalise economically unproductive farms and restore degraded farmland to productivity and health.[11] At Polyface Farms in Virginia, USA, regenerative agriculture practitioner Joel Salatin has transformed soil depth and habitat quality while building up an annual revenue of over two million dollars from sales of produce, making it an outstanding example of a **regenerative enterprise.**[12]

A straw mulch retains moisture, and at the same time projects the bare soil from the direct impacts of sun and rain, restricts weed growth and adds nutrients to the soil. KATY FOX/CELL

REVEGETATION

Removal of natural vegetation can be both a consequence of climate change and source of vulnerability to its effects; its restoration improves livelihoods, food security and physical resilience, at the same time creating a natural carbon sink

Land degradation is a major cause of climate change, and degraded soils are much more vulnerable to its impacts. While natural habitats vary in their resilience to climate change, simplification or removal of natural vegetation cover generally increases their vulnerability. Loss of vegetation reduces the landscape's ability to absorb and retain water, exacerbating the effects of floods and droughts. Bare soils are less able to withstand extremes of sun, wind and rain and more susceptible to erosion.

Revegetation is essential to soil protection and restoration, and is among the most effective ways known to sequester carbon

Deliberately restoring vegetation in degraded habitats by carefully selecting species that will improve soil structure and quality, or through farmer-assisted natural regeneration, is one of the most important ways permaculture promotes climate change adaptation. Restoring vegetation can both stabilise weather and increase resilience to climatic variability and uncertainty. Revegetation is essential to **soil protection and restoration**, and is among the most effective ways known to sequester carbon.

In Southern Malawi, climate change is exacerbating prior impacts of deforestation and land degradation resulting from inappropriate land use and agricultural policies which continue to promote synthetic fertilisers, pesticides and fungicides rather than **agroecological** methods. Forests are being systematically destroyed for new agricultural land, firewood and charcoal, and by the tobacco industry, which alone is responsible for 25 percent of forest clearance. As a result rainfall is increasingly late and erratic, causing crops to fail, homes and lives to be destroyed by flooding, and further eroding already severely damaged land. Most of these farmers, who make up 80 percent of the population, are caught in cycles of dependency upon synthetic agricultural products whose purchase leaves them in crippling debt. As a result, the average farmer's resilience to climate change is non-existent.

The African Moringa and Permaculture Project (AMPP, soon to merge with Kusamala[1]) is increasing resilience to climate change in both the local ecology and the livelihoods that depend on it by creating diverse and abundant Food Forests in and around numerous villages. In 2014 AMPP planted over 15,000 trees, with excellent survival rates. Food forests deliberately mimic natural ecological systems. Each forest is designed to include a high proportion of edible fruits and commercially valuable crops, such as ginger and passion fruit, which can hugely increase the quality and reliability of local livelihoods. The establishment of forest cover naturally contributes to **soil restoration**, water cycle regulation and many more ecosystem services. During the process of establishment, while the trees are still young, households can use these increasingly fertile spaces to grow vegetables for home and market. Not only do food forests support adaptation to climate change, as they develop they naturally sequester increasing quantities of carbon in trees and other plants, soils, and fungi growing on reserves of dead wood.

Farmer in Dowa district, Malawi. Climate Smart Agriculture for smallholder farmers in Malawi project. 2015 IGA GOZDOWSKA

Both AMPP and Kusamala also use agroforestry and **agroecology** techniques in staple crop fields to transition away from conventional forms of agriculture dependent on chemical inputs. Instead, farmers learn to use organic materials and manures, design permanent beds on contours and plant nitrogen fixing ground cover plants, shrubs and trees. These techniques use locally available cheap and/or free resources, naturally **retain water** (increasing resilience to erratic rainfall patterns), increase yields, improve soil structure and fertility, and at the same time sequester carbon. A combination of food forests and appropriate staple field designs can solve most, if not all, the hunger, malnutrition, poverty and environmental degradation problems commonly faced by Malawian farmers.

Agroforestry plot at Moringa, with ground cover and other crops growing beneath young trees. 2015 GA GOZDOWSKA

Seedlings at Kusamala Institute of Agriculture and Ecology, Lilongwe, Malawi. 2015 IGA GOZDOWSKA

4.4

AGRODIVERSITY

A buffer against natural environmental variability and source of adaptive capacity for small farmers everywhere, genetic diversity in crop plants and livestock is ever more vital in the face of climate change

Agrodiversity is the genetic diversity of cultivated plants and animals, whose use is vital in traditional smallholder farming worldwide. It refers to cultivated diversity that is both interspecific – among different species – and intraspecific – among multiple varieties of the same species. As different species and varieties have different properties and can tolerate or flourish in different conditions, access to agrodiversity allows farmers to adapt to changing conditions by focussing on whatever grows well, and maintain a genetic reservoir for adaptation to future change. Agrodiversity is a key component of **agroecology**: among other things it allows cultivation of polycultures, combinations of plants and animals that support each other's growth and exhibit high collective or systemic resilience to change.

Both traditional small-scale farmers and permaculturists are adapting to climate change by cultivating plants better suited to new conditions

Both traditional small-scale farmers and permaculturists are adapting to climate change by cultivating plants better suited to new conditions. In Southern Ghana, farmers cope with declining rainfall and rising temperatures by growing drought-tolerant and early-maturing varieties of staples, and increasing the range of secondary crops.[1] In North America, the Central Rocky Mountains Permaculture Institute experiments with a wide range of exotic species and varieties to find plants capable of producing food under changing climatic conditions.[2] It has established many of these as reliable domestic and commercial producers. Plants for a Future has compiled a database of over 7,000 edible and otherwise useful perennial plants, growing around 1,500 of these on its site in Cornwall.[3] Also in South West England, the Agroforestry Research Trust in Devon has an ongoing programme to identify and test exotic perennial plants for their performance in **agroecology** systems.[4]

Traditional polycultures have demonstrated benefits in addressing direct and indirect impacts of climate change, including water stress and disease. Experimental polycultures of sorghum, peanut and millet consistently yielded higher than the equivalent monocultures over a range of watering regimes; the differences in yield increased as water became more scarce.[5] Replacing monocultures of commercial rice hybrids with polycultures of hybrid and traditional varieties dramatically reduced incidences of fungal infection, eliminating needs for fungicide application, and almost doubled overall yields.[6]

Permaculture polycultures are less well-studied, with limited empirical data available. Preliminary field trials conducted by the UK Permaculture Association indicated slight increases in overall yields compared to monocultures, with a greater variety of edible produce available over a longer period of time.[7] Yields varied greatly from site to site, suggesting that planting conditions and growing techniques were also significant.

Seeds given away at a CELL event in Luxembourg, 2014. Reintroducing traditional and native seeds and creating networks for seed exchange are important techniques for restoring agrodiversity in places that have largely lost traditions of smallholder agriculture. KATY FOX/CELL

Fanny Chikukwa stands in his vegetable garden with his orchard behind him. He is the village headman for Stekete village and is a keen supporter of permaculture, using companion plantings of herbs and compost to assist his vegetable production. 2010 CHIKUKWA PROJECT/TERRENCE LEAHY

Terrace in orchard: Orchards slope down from the house and are terraced with swales to retain water and prevent erosion. 2014 CHIKUKWA PROJECT/TERRENCE LEAHY

Inside Orchard: Typical fruit trees include bananas, Mexican apple, guavas, citrus, avocados, pawpaws. Trees are planted in rows across the slope. 2014 CHIKUKWA PROJECT/TERRENCE LEAHY

Polycultural diversity and companion planting are themes of the Chikukwa project. Here a fig tree at the centre is festooned with passion fruit. CHIKUKWA PROJECT/TERRENCE LEAHY

4.5

AGROECOLOGY

Diverse, physically complex agro-ecosystems that emulate the structure and dynamics of natural systems remain productive and deliver ecosystem services under a wide range of weather conditions

Agroecological approaches maximise the number, diversity and quality of interrelationships among the organisms, species and populations at a site. **Agrodiversity** is usually a key dimension: in addition to using diverse species and varieties agroecological design locates these in both space and time in ways that maximise their beneficial interactions, including creation of **microclimates**. This strengthens ecological processes such as nutrient cycling, enhancement of soil life and biological suppression of pests and diseases.[1]

Agroecology is neither restricted nor original to permaculture, being characteristic of many traditional and indigenous farmings systems worldwide, some with histories extending centuries or even millenia.[2] Many of the common features of agroecological production reflect strategies farmers take to minimise exposure to risk of hunger due to crop failure in climates that are often marginal and/or unpredictable. This predisposes them for climate resilience in many different ways.[3]

Studies of recovery of farms in Nicaragua, Honduras and Guatemala after Hurricane Mitch in 1972 showed that compared to conventional farms, agroecological farms retained greater levels of topsoil, soil moisture and vegetation cover, and suffered lower levels of erosion and economic losses. The difference was higher the longer the land had been under agroecological production, and increased with increasing storm intensity.[4] In the 1992 droughts that severely impacted much of Southern Africa, female farmers in several districts of Zimbabwe mitigated the impact on food security through permaculture, **water management**, and use of **agrodiversity**, focusing on drought-tolerant crops.[5]

The Himalayan Permaculture Centre in Nepal supports farmer-led innovation, along with implementation of scientific research by establishments like Cornell University on sustainable rice intensification (SRI). Although focused on a single crop, SRI takes the agroecological approach of creating optimum systemic conditions for rice plants to flourish. It does this by giving careful attention to individual seedlings, promoting aerobic soil conditions for root growth and cultivating a conducive microbial ecology, and controlling weeds by mechanical methods that allow their incorporation into the soil.[6]

In the Food Forest, a 30-year-old permaculture site in Australia, sheep and geese graze between rows of trees, restricting growth of weeds. This eliminates any need for mechanical weeding or herbicide application, both of which are energy-intensive and polluting. It also allows retention of ground cover vegetation, which helps to reduce evaporation of water from the soil in this increasingly dry climate. The animals provide a range of additional products (meat, down, manure etc.), are more self-reliant in food, help clear the ground of fallen fruit, and can benefit from the shade and cooling effect of the trees.

Integrated assemblages of species, each of which benefits the others, seek to mimic factors promoting resilience and self-regulation in natural ecological systems. They provide a system that is more productive overall, includes buffers and safeguards against changing conditions for individual species, and at a systemic level is more resilient and adaptable to climate impacts than one dedicated to a narrower range of outputs. They also have unexpected emergent effects: the Food Forest initially provided a home for bettong, an endangered wild marsupial, purely for altruistic reasons. It later became apparent that they help with weed control by digging up and eating bulbs of the invasive exotic plant Oxalis, and with **revegetation** by burying seeds of native acacia trees, many of which later germinate.[7]

...forest gardens take the structure and ecological dynamics of a forest as a template for agroecological design

Best known in Europe, and to many people emblematic of temperate climate permaculture, forest gardens take the structure and ecological dynamics of a forest as a template for agroecological design.[8] Structurally, they include up to seven vertical layers, each made up of plants of similar size and habit. Dynamically, they mimic natural processes of succession by supporting the growth of different plants throughout the year. They provide seasonal **microclimates** for sun-loving plants in early spring, and for shade-tolerant plants at lower levels in summer once the trees are in full leaf. In addition to edible and other directly useful plants, some plants primarily fulfil ecological functions such as nitrogen fixation, attracting pollinators, and providing wildlife habitat, as well as human amenity value. Associations among cultivated plants and wildlife create emergent agroecological effects such as soil health, exchange of nutrients and pheromones among plants, and keeping pest and pathogen populations in check. An initial survey of newly-planted forest gardens in the UK reported that many had achieved success in overcoming adverse weather conditions, for example by planting windbreaks and creating warm **microclimates**. Almost all reported visible increases in biodiversity in many different taxa, including arthropods, reptiles, birds and annelids.[9]

A survey of newly-planted forest gardens in the UK reported that many had achieved success in overcoming adverse weather conditions

Bettong: an endangered marsupial; we decided to give them a home at the Food Forest, purely from an altruistic perspective, and then discovered that they dig up the bulbs of our worst weed (South African, Oxalis) and eat them. They have also assisted us with revegetation of the property by burying seeds of Acacia spp which germinate randomly and provide ongoing recruitment of young native trees. www.foodforest.com.au/fact-sheets/animals/bettongs

Above: Part of the agroecological design at the Food Forest in Australia, with vines, mulch, geese, cereal crops and a windbreak all working together in integrated fashion. GRAHAM BROOKMAN, FOOD FOREST

CAROLYN MONASTRA

Seed pods on Senna singueana *tree at Kusamala Institute of Agriculture & Ecology, Lilongwe, Malawi. 2015* IGA GOZDOWSKA

4.6

CREATION AND USE OF MICROCLIMATES

Designing in different conditions of temperature, humidity and exposure to the elements at small scales within a site allows a wider range of species to flourish, increasing diversity and adaptive capacity in the system overall

A key technique in **agroecology** and promotion of **agrodiversity** and is the identification and creation of microclimates: localised areas with distinctive conditions, for example of temperature, humidity, or exposure to sun or wind. Species and varieties that could not grow productively in the wider climate might flourish in a favourable microclimate, allowing cultivation of a wider range of crops. Microclimates also support distinctive communities of native species, increasing ecological diversity.

In the face of climate change, crops that become less well suited to new conditions may nonetheless be able to grow in specially created microclimates. Others previously restricted to specific microclimates may turn out to thrive in new conditions, allowing their cultivation on a wider scale. This is one way in which **agrodiversity** maintains a reservoir of crops with high collective resilience to climate change.

At Krameterhof, 1500 metres above sea level in the Austrian Alps, Sepp Holzer creates microclimates using many different techniques. Terraces and raised beds inclined towards the sun experience higher temperatures and greater intensity of sunlight. Curves in terraces, beds and paths vary this effect, creating multiple niches that favour particular species or ecological communities. Raised beds include buried organic matter that releases heat as it breaks down, elevating temperatures. Organic materials are also used as mulch, retaining water and heat at soil level. Carefully placed rocks act as reflectors and stores of heat. During the day, the south side of the rock acts as a suntrap where heat-loving plants can grow. The rock absorbs heat throughout the day and releases it at night, supporting plants that can not tolerate low temperatures. Locating rocks on the north side of ponds enhances these effects, as sunlight reflects onto the rock from the water surface. Techniques such as these allow cultivation of plants otherwise unknown at such altitudes and latitudes – including cherry, kiwi, sweet chestnut, apricots, grapes, and prickly pears – and prolong growing seasons for many others.[1]

The Central Rocky Mountains Permaculture Institute (CRMPI), at an altitude of 2200m in the state of Colorado, USA, actively experiments in montane agriculture as a source of resilience to climate change. Its highly variegated site encompasses wide natural differences in relief and aspect, enhanced by a design plan that uses built structures to create additional microclimates. This allows onsite cultivation of crop species and varieties from all over the world, a bank of **agrodiversity** to support adaptation to whatever climatic conditions come to prevail. As climate change makes current practices in commercial fruit production in adjacent lowland areas inviable, mountain agriculture using novel crops and techniques will become increasingly important for regional economic, livelihood and food security.[2]

CRMPI's greenhouses and indoor gardens house plants native to tropical, subtropical, desert and mediterranean climates, including multiple varieties of bananas, figs, grapes, apricots, apples, pears, plums, pomegranate, and edible cactus. They also extend growing seasons for temperate plants.[3]

Several technical measures retain internal temperatures conducive to plant health despite wide diurnal and seasonal variation outdoors. These include use of fan-driven heat exchangers known as climate batteries,[4] shading and venting to reduce direct heating, and exchange of heat with other household elements. Greenhouses thus form integral elements in an overall **bioclimatic architecture**.

In the increasingly hot and dry climate at the Food Forest in Australia, this mobile trough allows cattle to make the most of the precious shade under a carob tree.
FOOD FOREST/GRAHAM BROOKMAN

4.7

BIOCLIMATIC BUILDING

Designing buildings to fit with local climatic conditions and adapt to changing temperature regimes

Bioclimatic approaches harness natural processes to heat and cool buildings, through appropriate choice of materials, location and design. This allows maintenance of desired internal temperatures over a wide range of external conditions, future-proofing buildings against changes in weather patterns and possible temperature extremes. Such buildings are typically low-carbon or even carbon-negative in their operation: heating and cooling needs are low and can employ renewable sources, and many use natural building materials that are also carbon sinks.

Straw bale houses are a feature of permaculture projects in temperate and Mediterranean climates

Straw bale houses are a feature of permaculture projects in temperate and Mediterranean climates from Ireland and Northern England[1] to Italy.[2] Sieben Linden Ecovillage in Saxony houses most of its 140 or so residents in a series of multi-occupancy straw bale buildings.[3] Main facades and most windows face south to capture winter sun and minimise unwanted heat gains in summer and losses in winter. Supplementary heating comes from wood-fired stoves, most highly efficient in design. Research conducted by Kassel University showed that per capita GHG emissions associated with Sieben Linden's housing and heating are, respectively, ten percent and six percent of the German national average.[4] Wood for both heating and construction is cut from Sieben Linden's own forest as part of a long-term habitat enrichment programme, replacing extensive conifer plantations with more climate-resilient high-diversity native broadleaf woodlands.

The Central Rocky Mountains Permaculture Institute has developed innovative low-energy systems to regulate temperatures in the five large greenhouses onsite. These 'Climate Batteries' use fans to pump warm, moist daytime air underground where the water condenses, capturing both the thermal energy and the latent heat of condensation in the soil. This provides warm, moist soil conditions ideal for plant roots even at far lower above-ground temperatures. It also allows stored heat to return from the soil to the greenhouse interior in cooler weather.[5] The greenhouses form part of a wider bioclimatic design incorporating systems for heat exchange with adjacent buildings and thermal masses,

Bioclimatic straw bale residential building at Sieben Linden Ecovillage, with rooftop photovoltaic and solar thermal panels and main facade all facing south. 2012 TOM HENFREY

allowing greenhouses to be heated with residual heat from the sauna and living areas.

An integrated greenhouse and residential dwelling is also crucial to the bioclimatic design at Melliodora in Hepburn Springs, Australia.[6] The greenhouse acts as a temperature regulator for the house: providing an 'airlock' for one of the two main entrances, warming living spaces in winter and cooling them in summer. The house also incorporates substantial thermal mass: north-facing internal masonry structures that absorb the heat of the Southern Hemisphere winter sun and release it to the rest of the house. Partial shading by annual plants in the greenhouse and eaves over all north-facing structures reduces their summer exposure to the sun and so helps keep the house cool.

Two biodigestor toilets that convert human waste into fuel for cooking in the kitchen at Rancho Mastatal, Mastatal, Costa Rica, 2011
CAROLYN MONASTRA

4.8

ENERGY DESCENT

Local communities across the world are using permaculture to design strategies to build resilience to climate change and expected declines in availability of cheap energy

Permaculture is a vital tool in strategies for progressive reduction of energy inputs necessary for economic activity while simultaneously increasing prosperity.[1] It challenges established macroeconomic theory and associated policy measures that assume provision of wealth and wellbeing to rely on continued economic growth, which has never genuinely been decoupled from increases in carbon emissions and other forms of environmental damage.[2] By freeing policy from assumptions incompatible with effective action towards mitigation, it also broadens the range of policy approaches available to support adaptation. By finding solutions based on the convergence of needs, ethics and available resources rather than theoretical dogma, permaculture can support diverse, intersecting forms of low-carbon, zero-carbon and carbon-negative economic activity.[3]

When self-organised community action is the main strategy for energy descent it increases social capital, providing a powerful basis for long-term adaptation to climate change

The Transition approach to **Economic Localisation** is one example of this. It emerged when permaculture students at Kinsale College in Ireland completed an end-of-course project on designing a community-based strategy for coping with peak oil.[4] The movement that developed on the basis of their insights quickly adopted climate change as a second key driver.[5] Transition groups apply permaculture to designing out the causes of climate change at the same time as designing in desirable outcomes. This allows identification of low-energy pathways to support well-being at the scale of the local community.[6]

When self-organised community action is the main strategy for energy descent it increases social capital, providing a powerful basis for long-term adaptation to climate change. The Transition Streets project in Totnes, South West England, encouraged small groups of immediate neighbours to meet and share ideas and concerns about climate change. This achieved highly cost-effective reductions in household carbon emissions through behaviour change, energy efficiency measures, and renewable energy installation.[7] Independent evaluation by the UK Government suggested that intangible social benefits were equally important: by coming together in this way, neighbours got to know each other far better than they had done before.[8] Social capital of this type is an important resource for climate change adaptation, supporting people's ability to cooperate in the face of any crisis.

When Canterbury, New Zealand suffered severe earthquakes in 2010 and 2011, relief and reconstruction efforts benefited greatly from prior work by Project Lyttleton, a local initiative dedicated to building community resilience.[9] The group ran a time bank, a classic example of **social technology** through which residents exchange labour and skills on a peer-to-peer basis. After the earthquake, the time-bank became a vital part of the relief effort, rapidly mobilising people to check on elderly residents, providing childcare, undertaking minor household repairs, and delivering meals to vulnerable people. Project Lyttleton's work both complemented that of the authorities and challenged many of their ideas and perceptions. This type of contribution to **changing worldviews** becomes increasingly important as established understandings and practices become less and less relevant in the face of climate change.[10] Transition has been successful in improving conditions in very poor areas, where it provides a model for responses to economic hardship that may arise from climate change. Brasilândia, a very low income suburb in São Paulo, Brasil, has hosted a Transition Initiative since 2010.[11] It ranks low in all conventional development measures, with the second-lowest Human Development Index in the city and a zero score in the official São Paulo City Observatory report on culture. However, Ir BEM São Paulo, a city-wide survey of neighbourhoods conducted since 2009, shows steady improvements in all areas of Transicion Brasilândia's work, including frequency of cultural events, standards of health care, and quality of community relations and civic responsibility. Indices of participation in voluntary activity, awareness of environmental impacts of consumer goods, level of community ownership, and peaceful co-existence of different religious groups, have all risen to the highest in the city.[12]

Posters produced by Transition Minett in Luxembourg reflect different personal perspectives on energy descent. NORRY SCHNEIDER

4.9

BIOREGIONALISM AND ECONOMIC LOCALISATION

Grounding local and regional economies in prevailing ecological, social and cultural realities and creating appropriate wider linkages to build flexible and resilient systems for meeting human needs

Bioregionalism is the notion that human societies and economies organise primarily in relation to the specific ecological and cultural details of place.[1] Arising as a philosophy and movement in North America during the 1970s, it has developed initially in parallel and later in synergy with permaculture, which has become among its most important practical tools.[2] Both view climate change as a consequence of dislocation between the causes and consequences of human environmental impacts, and restoration of these a vital part of any remedy.

A key implication of bioregionalism is economic localisation: meeting material needs as far as possible using resources available within the bioregions.[3] The consequences are thus directly visible to and felt by the users themselves – in contrast to the dislocation of cause and effect that is one of the key causes of climate change. Neither bioregionalism or localisation are isolationist movements: both are about improving the quality of relationships, between people and the ecological setting in which they live, among people within a locality, among localities within and across bioregions, and among bioregion, nations and at all scales. A basic thesis is that local economies, self-sufficient as far as possible, can enter into trade and other forms of interaction based on values of sharing, accountability and cooperation.[4]

Such a combination of local self-reliance and cooperative interchange at wider scales is or has been a source of resilience and adaptability in many indigenous and traditional societies. Traditional residents of frost-prone areas of the New Guinea Highlands relied on fields close to home in normal circumstances, kept secondary fields in different microclimatic zones to safeguard against seasonal frosts likely to damage one field but not the other, and maintained intricate social support networks to allow structured temporary migration to lowland areas when occasional extreme frosts totally wiped out agricultural production in any given year.[5] In Bristol, South West England, mutual support among small scale independent producers using permaculture and other low

Integrated Ecovillage Design Education and Transition Towns training to build the capacity of indigenous people, especially women, in the Laxmipur Block of Koraput District to increase food security, build social cohesion and address climate change. THREAD

input approaches allowed those worst affected by floods in 2013 to continue production. This informal social permaculture network also connects with projects in the wider bioregion and geographical region through trade, financial investment, training, and exchange of materials, skills and services, many of these vital to the survival of rural projects.

Economic localisation is a key strategy of groups in the Transition Movement, a network of several thousand local initiatives in over 40 countries worldwide using permaculture to devise and implement community-led responses to climate change.[6] By identifying and creating alternatives to carbon-intensive and climate-vulnerable features of local economies, they devise resilience-building strategies that go beyond the mitigation-adaptation distinction.[7] Local Economic Blueprints developed by several Transition Initiatives show massive potential financial benefits of partial localisation of key economic sectors such as food production, energy, and housing.[8] Localisation thus directs money and other resources towards **energy descent**, creating new entrepreneurial opportunities based on **regenerative enterprise**.

Smokeless stove, Himla, made entirely from local resources. CHRIS EVANS/ HIMALAYAN PERMACULTURE CENTRE

REGENERATIVE ENTERPRISE

A view of enterprise as a means to regenerate natural, social, cultural and infrastructural systems as well as growing financial capital: both more adaptable to the consequences of climate change and able to support wider adaptation

The direct economic impacts of climate change include the costs of adaptation and mitigation. Indirect impacts arise from the deeper need to restructure economic systems requiring continual inputs of large quantities of fossil fuel energy (and hence high outputs of greenhouse gases) and externalisation of environmental and social costs.[1] Within a global economy structurally dependent on growth, and with no way to decouple growth from rising greenhouse gas emissions, the 'green economy' can remain at best an adjunct to business as usual.[2] Even without taking social costs into account, research conducted within the United Nations Environmental Programme has shown that if environmental damage was accounted for, none of the major global industries would be cost-effective.[3] Accordingly, permaculturalists have sought alternative ways of doing business that neither directly contribute to climate change nor undermine the social and natural capital necessary for adaptation.

Research conducted within the United Nations Environmental Programme has shown that if environmental damage was accounted for, none of the major global industries would be cost-effective

The Reconomy project is an important feature of the Transition movement's efforts towards **economic localisation**. By taking an entrepreneurial approach to community-based responses to climate change, it seeks to make them financially sustainable. A report from the UK included examples of community-run energy cooperatives, bus services, bakeries, cafés, pubs, bike workshops, health care providers, builders, housing providers, local currencies and banks.[4] Longer term, and especially when co-existing in the same area, such Transition Enterprises can be the basis for local economies flexible enough to sustain provision of material needs through the economic restructuring involved in **energy descent**. Many Transition Enterprises take a cooperative approach, and there is strong evidence that this makes sense in relation to adaptation. Analytically, the seven principles of cooperative enterprise map closely onto the properties of resilient systems with high adaptive capacity. In places with long-standing traditions of cooperative business and established networks of mature enterprises, trading among each other and supporting the creation and growth of new co-ops, the result is highly resilient regional economies able to respond flexibly to changing conditions.[5] Committed to a geographical place, such businesses and networks provide a powerful basis for adaptation to the economic impacts of climate change: a compelling example of the practical benefits of **commons-based governance**.

The concept of Regenerative Enterprise deepens these insights through the systematic application of permaculture design.[6] In addition to the familiar concept of financial capital, it recognises seven other forms of capital: material, living, social, cultural, intellectual, experiential and spiritual. All eight forms of capital are essential to a healthy economy and society, and to effective climate adaptation. Regenerative enterprises seek to cultivate rather than extract capital, focusing on the quality and interconnection of the capital they create, not just quantity, and so reverse the consequences of narrow focus on financial capital. They exist in collaborative relationships within enterprise ecologies in which different enterprises take responsibility for nurturing different forms of capital. Regenerative Enterprise is a key influence on the wider theory of Regenerative Capitalism: an approach to economics based on biomimicry, consistent (unlike conventional economics) with contemporary science and hence capable of supporting long-term economic and social flourishing.[7]

An example of a regenerative enterprise is Nova Mondo, which harnesses the potential of cacao to support cultural and ecological regeneration: supporting ecologically sound farming practices and ethical business among supplier partners in Ecuador, Nicaragua and Belize and investing surplus revenues in rainforest protection.[8] Another is North American firm Guayakí, which works in partnership with Aché Guayakí people and other groups indigenous to the South American Atlantic forest to support their economic self-reliance and consequently their cultural self-determination. It markets yerba mate grown in the shade of the forest canopy by Aché cultivators and offers various forms of technical and social support.[9] It invest profits in replanting native hardwoods in the Atlantic Forest, a highly degraded and threatened habitat.

Figure 2: Eight Forms of Capital

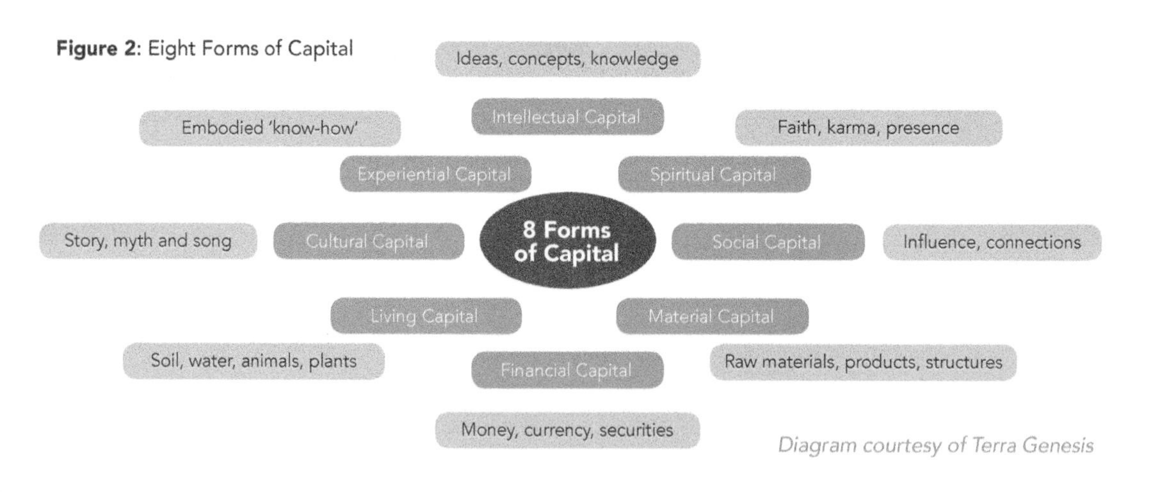

Diagram courtesy of Terra Genesis

Lush North America's first Direct Trade purchase.

International Cosmetics firm Lush has become a leading example of Regenerative Enterprise through its ongoing efforts to reconfigure its production processes according to permaculture design principles.

Lush actively seeks to source raw ingredients from permaculture-based producers, and in doing so often supports them to develop their operations to the necessary standard and scale. Here, buyers for Lush North America sign a contract with cacao producers, then get hands on helping to restore degraded habitats as multifunctional polycultures, including the commercial important cacao trees.

TERRA GENESIS/GREGOY LANDUA

Preparing the earth to plant peanuts at Verde Energia, a permaculture community in Lanas, Costa Rica, 2011
CAROLYN MONASTRA

4.11

COMMONS-BASED GOVERNANCE

Flexible, responsive systems for collective ownership, decision-making and allocation of rights and responsibilities

Commons are collectively owned resources, collaboratively managed by their co-users under formal or informal governance systems.[1] Their management institutions are finely attuned to the specifics of place and closely link ecological and social reality. For this reason, traditional commons are a crucial ingredient in all documented cases where human groups self-organise to achieve sustainability and/or resilience.[2] Creation of new commons has been mooted as a central strategy in climate mitigation and adaptation.[3] Permaculture is one of many social movements dedicated to the creation of new commons, and along with these novel institutional mechanisms for climate adaptation that transcend the limitations of state and market.[4]

> Much agrodiversity is maintained and circulated through informal projects and networks for propagating and exchanging plants

Like the species and ecological formations that inspire permaculture design, well-governed commons are resilient, self-limiting, adaptive entities that promote regeneration of landscapes and social systems. Nobel Laureate Elinor Ostrom's research on commons identified five requirements for adaptive governance.[5] These effectively translate into a set of baseline social conditions for successful adaptation to climate change. They are:

- Access of users to accurate and relevant information;
- **Social technologies** to enforce user responsibility and compliance with management rules;
- Effective mechanisms for **conflict transformation**;
- Flexible infrastructure for both internal operations and external links; and
- Encouraging adaptation in the face of changing external circumstances.

The new commons created as part of permaculture are diverse in nature and form. Much **agrodiversity** is maintained and circulated through informal projects and networks for propagating and exchanging plants. **Popular education** creates knowledge commons through which concepts such as permaculture principles and design methods, and theories around **economic localisation** and **energy descent** are developed, tested and refined. Knowledge commons around **social technologies** and tools for **personal resilience** are created, refined and shared through their use in meetings, workshops, events, training exercises, courses, and other forms of group work. Many permaculture projects own and manage their physical base as some form of commons through legal structures such as trusts, cooperatives and community interest companies that support inclusive decision-making processes.

The Centre for Ecological Learning Luxembourg (CELL) is a national and regional hub for permaculture and Transition, legally structured as a non-profit organisation. It operates as a commons and in addition supports establishment of new organisations for creation and management of common pool resources.[6] Relationships with related organisations are cooperative in nature, enacting an ecological model in which each helps create conditions for the success of others. CELL is currently developing design and consultancy services as an income-generating **regenerative enterprise**. Resulting financial surpluses will be redistributed through its organisational ecology towards citizen groups whose activities do not generate income. This is one way in which CELL cultivates conditions for the emergence of a constellation of organisations with diverse legal forms, pursuing distinct specific objectives within a common aim of promoting societal and environmental health through citizen-led responses to climate change.

A key enterprise that has emerged from CELL is TERRA (Transition and Education for a Resilient and Regenerative Agriculture), a food-growing, educational and community-building project founded in 2014.[7] TERRA operates a form of Community Supported Agriculture (equivalent to AMAPs in France and Solidarlandwirtschaft in Germany) a relatively new model of food production and distribution that rests on cooperative legal forms and seeks to blur the distinction between consumer and producer. TERRA operates as a cooperative society whose 150 members act as co-owners and co-managers. It supports three paid employees whose work constitutes an ongoing programme of action learning on **soil regeneration, agroecology, popular education,** and use of **social technologies**: running the farm, distributing weekly produce shares to members, and organising learning events and seasonal celebrations. After only one year TERRA had

CELL/TERRA Permaculture Course PEACE ADVOCATE PHOTOGRAPHY

increased availability of locally produced organic food and achieved measurable improvements in soil quality, showing concrete progress towards its long-term aim of combining physical and social techniques to help build a more climate-resilient food system.

Another example of a commons-based system of governance inspired by permaculture design is Biovilla, a Portuguese sustainability cooperative, founded in 2010 and focused on nature tourism, education for sustainability and landscape regeneration.[8] Biovilla is moving towards a sociocratic governance model based on the horizontal concepts of zoning, closed-cycles, interdependency, systems thinking and possibility management. This allows it to have a flexible, transparent and constantly adapting organizational ecosystem (Figure 3). Each member is considered to have an equal share in the social capital and all members take part in all major decisions. New technological solutions such as WhatsApp support these inclusive processes, helping groups to stay connected.

Biovilla seeks to demonstrate the concept of tri-dimensional alignment within an organisation. This proposes that true sustainability can only emerge when the legal framework is fully aligned with the governance system, and in turn with decision-making processes and finally the organisation's principles, values and mission (Figure 4). So managing the commons is about more than cooperative ownership: it also implies democratic, transparent, engaging and participatory management of the collective means of production.

Figure 3: Nested scales of governance in Biovilla

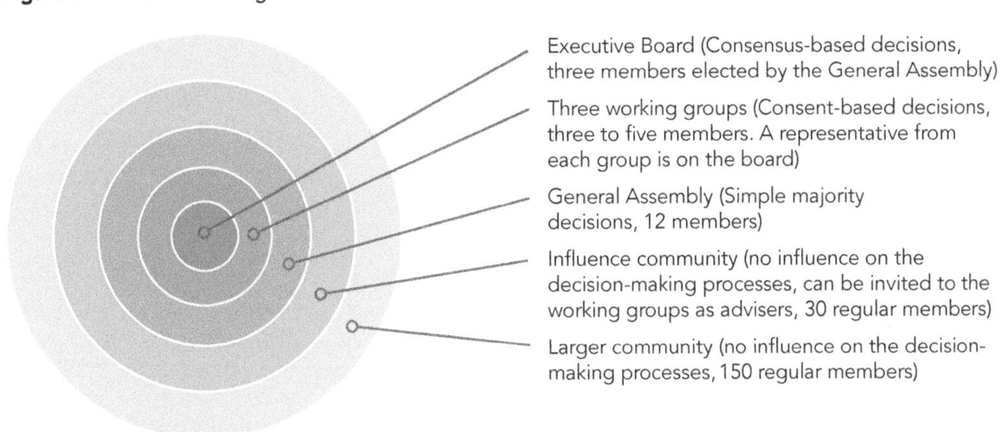

Figure 4: Tridimensional Organisational Alignment: a prerequisite for sustainability

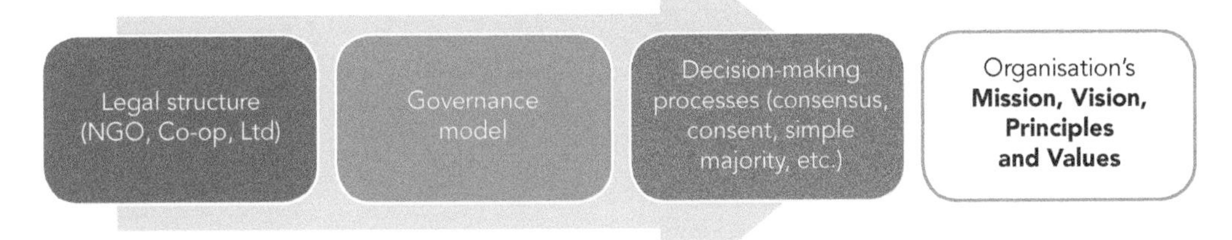

Designing a large property
KATY FOX/CELL

4.12

SOCIAL TECHNOLOGIES

Imagining, creating and employing the new social structures and processes necessary for adapting to and living with climate change

Permaculture approaches recognise that effective responses to climate change require social as well as material change. They use a range of social technologies in order to locate physical interventions in communities of practice at all levels, and to embed these in their wider social contexts. Social Permaculture translates ecological principles to social situations in order to foster flexible and creative thought and action, nurturing the collective intelligence of groups such as local residents, stakeholders to a project, or in educational settings.[1] Its applications are especially well developed in Transition and other movements for **energy descent** and **economic localisation**, in **popular education** and **commons-based governance**, and are an important feature of many cases of **regenerative enterprise**.

Social technologies can contribute to the design, maintenance and success of innovative, low-impact projects at local and regional levels

Social technologies can contribute to the design, maintenance and success of innovative, low-impact projects at local and regional levels. In order to ensure respectful, inclusive, constructive participation of all involved stakeholders, intervention processes need to be planned strategically and carefully, designed with attention to inclusion and participation, depending on the objective of the intervention, and seen through with skilled facilitation.

How the decisions in a community group are taken affects how people in this group will feel about how much or how little their voice is being heard. In an active shift from representative to participatory democracy, permaculture projects, ecovillages and the Transition movement have appropriated, developed and refined methods to include as many people as possible, keep processes moving, enable **conflict transformation**, use appropriate communication, create community contracts and ensure accountability, design and implement governance and decision-making processes, develop networks, and ensure cohesion through conviviality, creativity and fun. Social technologies have also been applied to healing existing divisions in communities: the Chikukwa Project in Zimbabwe has set up female empowerment groups, and village talking circles to address stigma against sufferers of HIV and AIDS.[2]

The Eco-Social Matrix (ESM) developed by permaculture teacher Robyn Francis explores the interplay of natural, social and infrastructural factors to assess how land use decisions impact the environment and people's lives, particularly how they access resources and meet their needs.[3] The matrix can be applied to an individual property, a neighbourhood, a bioregion or any larger scale. It has been especially useful in helping ecovillages and other permaculture projects that had been rather inward-looking integrate more successful and productively with neighbouring communities. The New South Wales Department of planning has adopted it as a 'Catchment Planning Framework' within its guidelines for rural residential development.[4]

Social technologies are fundamental to the work of the Centre for Ecological Learning Luxembourg (CELL), a national and regional hub for permaculture and Transition.[5] CELL raises awareness and designs regenerative systems for changemakers and grassroots groups interested in bioregional resilience, in order to empower, nurture and catalyse connections between living beings and the places they inhabit. It uses a range of social technologies to ensure that all its activities embody its core values of mindfulness, coherence, authenticity, creativity and celebration.

CELL's members worked collaboratively to elaborate an initial mission based on permaculture ethics and principles, and to devise operations that can support learning of the type necessary for constructive action on climate change adaptation. Governance employs Sociocracy, an established process for self-organisation, distributed authority and inclusive decision-making within a group.[6]

Ethics of inclusion and equality are designed into processes at all levels, including founding documents, strategy, meetings and trainings, project development, and collaboration with a wider constellation of stakeholders and organisations. All of these use methods that enable meaningful participation and expression of creativity. These methods include Systemic Consenting, a simple way for a community collectively to determine the least favourable among available options,[7] and Open Space technology, a method based around self-organisation and the principle that the stakeholders involved will build their own agenda according to their needs.[8] Since its foundation in 2011, CELL has in this way fostered emergence of an ecology of interconnected groups, organisations, projects and initiatives that provides a powerful and basis for proactive and reactive responses to uncertain future impacts of climate change.

Collaborative Learning
KATY FOX/CELL

4.13

CONFLICT TRANSFORMATION

Conflicts are inevitable in the face of climate change; when managed creatively they can become opportunities for improved understanding, new forms of collaboration and increased social cohesion

Climate-related conflicts are already numerous and likely to increase in frequency, magnitude and severity as the impacts of climate change get worse. They can arise over issues like access to increasingly scarce water sources or reliably productive soil, or over food shortages, and can both provoke and escalate in response to migration.

Conflict transformation can help address disputes arising directly from either climate change itself or from adaptation efforts

Often they are exacerbated by inequalities, both in exposure to impacts and access to financial resources that can buffer these.[1] The influence on adaptation policies and strategies of security-driven agendas, often reflecting military preoccupations, worsens this still: addressing symptoms rather than causes and treating victims of climate change as threats to the physical and livelihood security of others.

Among the most important **social technologies** used by many permaculture projects are conflict transformation mechanisms. These can help address disputes arising directly from either climate change itself or from adaptation efforts. They can also heal existing schisms among and within communities, creating a more solid social base for collective efforts at adaptation.

The Chikukwa Project in Zimbabwe has trained around 50 villagers in conflict transformation. It has developed its own system that combines traditional social technologies with established tools from European countercultural movements, setting up groups for Building Constructive Community Relations (BCCR). When people cut trees planted on a wooded ridge to ensure **water retention** in the landscape, it caused soil erosion on some downstream farms, silting on others. A two-day workshop organised by the local BCCR group led to agreement and implementation of an action plan in which all those affected took part amicably: replanting cut areas and extending the planted area overall, and building new dykes and levees in the stream.[2]

Tamera Ecovillage in Portugal is home to the Tamera Peace University, which offers courses on sustainable cultures of peace and follow-up workshops on community-building. Peace pilgrimages to areas like Colombia and Palestine link efforts to build harmonious communities locally with contributions to overcoming conflict and its consequences more widely, supporting reconciliation and forging lasting friendships and collaborations.[3]

Many permaculture projects in areas afflicted by deep cultural and ethnic conflicts build responses to these into their social dimensions. Hava and Adam Eco-Educational Farm actively works to overcome separation and build understanding and collaboration between Israelis and Palestinians. This has already fed into practical work to support climate change adaptation: Palestinian farmers have drawn on their traditional knowledge to introduce traditional irrigation techniques and planting practices, and provide seeds of drought-tolerant native crops. Use of permaculture techniques to grow more in smaller spaces has helped farmers to respond to scarcity of land.[4] This is one of several permaculture projects in Israel and Palestine who employ permaculture as a common language to support reconciliation and shared action. Such activities have provoked repression by the Israeli army, which in November 2000 reportedly raided, sacked and prohibited access to a permaculture centre in the West Bank village of Marda.[5]

Beyond Israel and Palestine, Los Angeles Eco-Village in California was founded as part of community rebuilding efforts in Wilshire/Koreatown, a highly ethnically diverse neighbourhood that suffered great loss of life and physical damage during civil unrest sparked by institutional racism in 1992. Elsewhere in the USA, Growing Power in Milwaukee addresses what its founder Will Allen describes as food racism: the status of many African-American and Latino neighbourhoods across the country as 'food deserts'. Its urban farms, distribution hubs and retail outlets are the only sources of fresh, nutritious produce for most residents of an area where nutrition-related health problems are endemic.[6] In addition to their direct benefits, such initiatives raise the prospect that future responses to climate change arise from more cohesive communities, drawing on their full range of human diversity, and less afflicted by existing schisms that climate change impacts might exacerbate.

Conflict transformation workshop at Chikukwa. The CELUCT project has set up a programme of conflict transformation to mediate disputes in the community. Here, women at a workshop watch a roleplay of a conflict staged by their Conflict Transformation committee. 2010 TERRENCE LEAHY/ CHIKUKWA PROJECT

4.14

PERSONAL RESILIENCE

Many tools now exist for coping with the personal challenges that come from accepting the reality and full implications of climate change

Fully taking on board the implications of climate change can be so challenging to established assumptions and worldviews that it is experienced as a threat to personal integrity or even survival.[1] This is true both of its direct consequences and the extent of the social and economic changes that successful adaptation to climate change in a carbon-constrained world will entail.[2]

Making climate change a mythic context for understanding the human place in the world provides a source of hope

From development of practical tools for addressing this, a sophisticated applied psychology of climate change has emerged among groups taking permaculture-based action on climate change.

Permaculture's holistic perspective and orientation towards solutions are equally useful for personal and group work as they are for practical action.[3] Viewing the self in ecological perspective implies attention to relationships, both interpersonal and with the material world.[4] Numerous tools for personal and group development exist; permaculture provides an integrated framework within which to select and apply these in the most effective way for any particular situation.[5]

This has been best developed in the Transition movement, whose strong focus on climate change makes the need for attention to personal resilience impossible to ignore.[6] Inner Transition addresses the need to cope with environmental and social challenges that are now unavoidable, and the potential for personal empowerment that comes from taking responsibility for addressing these.[7] Making climate change a mythic context for understanding the human place in the world provides a source of hope.[8] Taking action to adapt in ways that have positive environmental and social consequences makes this hope meaningful and productive.[9] With no fixed set of methods in place, individuals employ many different specific philosophies and practices. Joanna Macy's 'Work that Reconnects' aims to heal the sense of separation from the natural world, the

Earth, and from each other, acknowledging and expressing the pain of climate change and other forms of environmental damage and transforming them into motivation for positive action.[10] Many people active in Transition and permaculture are also involved in Quakerism, whose concept of 'Right Relationship' underpins sophisticated philosophical analysis of the meaning of climate change and necessity for constructive responses.[11]

Data on the personal resilience of people actively involved in Transition reveal a complex picture: some come to the movement seeking a source of personal resilience, others because their prior resilience gives them strength to act.[12] Research on founding members of Transition and other grassroots groups taking action on climate change shows how their work creates what are termed salutogenic environments: settings that allow people to react to current issues in ways that promote their emotional health by making sense of them, supporting constructive and achievable action, and finding shared meaning in this work.[13] In this way, climate change becomes less a threat to humanity, more an invitation to global society to step up to a new level of maturity and collective responsibility.

Planting seedlings as part of the Force Majeure project, Sagehen Creek Field Station, Truckee, California, USA 2014
CAROLYN MONASTRA

4.15

CHANGING WORLDVIEWS

As established worldviews become redundant or dangerous, new alternatives are emerging, better suited to present-day realities

Climate change is, as much as anything else, a conceptual challenge. As a 'wicked problem', it is not amenable to simple cause-and-effect analysis or resolution of the type that underlies conventional policy measures.[1] It is better viewed not as a problem in itself, but a persistent condition that sets the context in which all other problems must now be addressed.[2] Understanding this is not straightforward to minds socialised in Euro-American cultures. For them, climate change is easy to understand in the abstract, but almost impossible to internalise and link to the realities of personal, public and political life in ways that lead to meaningful action.[3]

Many experienced permaculture practitioners consider themselves to experience the world in ways radically different from the Cartesian model.[4] A holistic, integrated view, permaculture emphasises patterns and processes, relationships and interconnectedness rather than things and discrete events. Many of the principles attributed to founder Bill Mollison come across like Zen koans, designed to disrupt established patterns of thinking: "Everything Gardens" or "The Problem is the Solution". The twelve design principles described by co-founder David Holmgren[5] superficially appear less abstract, and can easily be described and learnt in an hour or so, but reveal their full depth only through years or even decades of sustained application.

In practice, this manifests in various ways. Where permaculture works in connection with **traditional knowledge**, it often incorporates elements of traditional belief. At Chikukwa in Zimbabwe, adoption of permaculture as a land management and livelihood diversification strategy has helped revitalise traditional beliefs around care for the land, with village leaders keen to emphasise the ritual aspects of this.[6] The Mesoamerican Permaculture Institute in Guatemala teaches and applies permaculture within a framework based on Maya cosmology.[7] The Sarvodaya movement in Sri Lanka finds a productive synergy between permaculture ethics and Buddhist commitments to caring for nature. Kibbutz Lotan in Israel employs *tikkun olam*, a Hebrew term for the healing of the world, within an eco-socialist interpretation based on teachings about the environment in the Torah and other Jewish texts.[8] Russian permaculture, ecovillage and back-to-the-land movements are heavily inspired by the Ringing Cedars series of books by Vladimir Megré and their main character Anastasia, a young woman who dwells deep in the forest in harmony with nature and espouses a radical philosophy of personal growth through nature connection.[9]

Many prominent permaculturists, including Australian permaculture teachers Rosemary Morrow and Robin Clayfield, see their work as closely related to Earth-based personal spiritualities.[10] North American writer and environmental activist Starhawk explicitly links permaculture and nature mysticism in a spiritual programme to support personal action on climate change and other environmental issues.[11] This has much in common with Inner Transition, the tools for supporting **personal resilience** in the face of climate change developed within the Transition Movement.

Starhawk has raised the provocative – and entirely reasonable – question of what shape global climate change adaptation and mitigation measures would take if the three permaculture ethics (in her words, Care for the Earth, Care for People, Care for the Future) were the basis of policy and law.[12] In her view this would imply legal, policy and fiscal measures to support regenerative enterprise; legal measures to prevent environmental and social damage by profit-seeking corporations; basing business accounting on contributions to **energy descent** rather than fiscal returns; enterprises rooted in place and hence supporting **bioregionalism and economic localisation**; incentivising careers and livelihoods that enhance environmental and social care; and massive programmes of **popular education** to support all of these transitions. These structural measures would in turn support practical outcomes similar to those reported in these sections, but on a far deeper and wider scale. Many such visions are possible; what this one shows is that effective practical action and the policy measures necessary to achieve this depend on worldviews very different from those on which present global political and economic systems are based.

The temple of Eco Yoga Park, General Rodriguez, Argentina CAROLYN MONASTRA

4.16

INDIGENOUS AND LOCAL KNOWLEDGE

Traditional and local knowledge often inform permaculture-based responses to climate change; at the same time, many indigenous groups have incorporated permaculture into their own adaptation strategies

As Chapter 3.1 describes, the traditional environmental knowledge of many indigenous peoples includes mechanisms for navigating the variability, uncertainty and unpredictability inherent in complex ecological systems. This can predispose them to adapt successfully to the effects of climate change. However, these effects, especially when they are accompanied by or coincide with social, cultural, economic and ecological change, may have impacts beyond what traditional systems can accommodate. In many places, much important knowledge has been lost, or the socio-cultural systems that support it changed dramatically. Permaculture can support the recovery of traditional knowledge, its integration with other forms of knowledge and practice, and the development of new approaches viable in present-day circumstances. Key examples include the Chikukwa Project in Zimbabwe, extensively covered in other chapters in this section, the Mesoamerican Permaculture Institute (IMAP) in Guatemala, and The Suvraga Aguyt Cooperative in Mongolia.

IMAP was founded in 2000 by a group of Kakchiquel Maya concerned about the serious environmental, social and cultural problems affecting Guatemala. Its approach blends permaculture with traditional Mayan knowledge, practices, beliefs and cosmology. Through community service, **popular education**, consultancy and group visits, it supports local indigenous farmers to respond to key climate change impacts such as drought and excess rainfall, loss of soil fertility and soil erosion, and increasing pests and diseases. Courses and workshops combine permaculture principles and design methods; novel techniques in areas like **water management**, **soil protection** and use of **social technologies** for community organisation; and aspects of tradition belief and practice like the Mayan planting calendar. Participants in courses include community leaders and representatives of NGOs from throughout Guatemala. Most transmit their learning further through word of mouth and informal peer learning among farmers, so new knowledge remains tightly embedded in traditional systems.

A particularly important areas of IMAP's work is the conservation and use of traditional seed varieties: historically selected to fit local conditions and needs, and representing a

Rony Lec, Coordinator and Co-Founder of the Mesoamerican Permaculture Institute, explains the Mayan planting calendar during a permaculture design certificate course. ZIPPORAH LOMAX

reservoir of **agrodiversity** that can underpin flexible production strategies and resilience to changing weather conditions. As in many other places, traditional varieties in Guatemala are themselves under great threat due to aggressive marketing of commercial hybrid seeds by large agricultural suppliers. Less suited to local conditions and with little tolerance for variability in growing conditions, they are not integrated with traditional production systems and techniques. Adoption of hybrid seeds can contribute to degradation of local knowledge and capacity for self-sufficiency, permanently locking people into dependency upon both the seeds and the high levels of external inputs (fertilisers, pesticides, etc.) required for their cultivation. It thus undermines local resilience and adaptive capacity at a systemic level, so people are more likely to require external assistance to support adaptation.

IMAP's contribution to disaster relief efforts following hurricanes in 2005, 2008 and 2010 highlights how strong local capacities for risk management and response can complement centralised efforts on the part of governments and international agencies. Use of permaculture methods to support temporary provision of food, water and shelter for destitute families created a model, refined on each occasion, on which to base future responses. Partnerships with international and local relief services and NGOs established during this work provide a platform for future collaboration. By lobbying, developing educational resources and implementing measures to control soil

Amazonian Shipibo Community work to replant the rainforest (Zone 2-3). GREGORY LANDUA/TERRA-GENESIS

erosion, IMAP is building longer term measures that will directly mitigate the effects of future extreme weather events and strengthen local capacities to cope with and recover from them.

The Suvraga Aguyt Cooperative in Mongolia uses permaculture as a framework to combine traditional and novel expertise to adapt to changing weather conditions, including winter temperatures as low as -40°C and scarce and unpredictable rainfall, that make their traditional way of life inviable. Livestock shelters have been redesigned for heat retention, and are now heated with animal manure and insulated with furs that in previous times were discarded. People with little or no previous experience of agriculture are now cultivating a range of vegetables for home consumption and sale, broadening their livelihood base. Abandoned Soviet era buildings have been revamped as passive solar greenhouses, creating favourable **microclimates** for seedling and plant growth so a greater range of crops can grow.[1]

The Sarvodaya movement in Sri Lanka hosts annual Permaculture Design courses that support residents to supplement their existing agricultural knowledge with suitable, culturally compatible, permaculture techniques.[2] Permaculture ethics intersect closely with villagers' Buddhist values that revere all life, an example of the links between permaculture practice and **changing worldviews**.

Watering the crops, Eco Yoga Park, General Rodriguez, Argentina 2011 CAROLYN MONASTRA

4.17

POPULAR EDUCATION

Through grassroots processes of informal, non-formal, and – at times – formal education, permaculture has created a global action learning community able to innovate and spread new ideas and practices for climate change adaptation

Adapting to both the direct effects of climate change and the economic, social and cultural consequences of massive reductions in global levels of carbon emissions will require collective learning on an unprecedented scale. Permaculture has grown and spread worldwide as a popular education movement. In the process it has incorporated new ideas and practices as it adjusts to different settings and changing circumstances.

At the heart of permaculture practice are processes of action learning – trying something out, observing the effects, and adjusting future practice accordingly. Teaching takes a similar approach: open and pragmatic, valuing and drawing on learners' pre-existing knowledge.[1] Permaculture is taught not as orthodoxy, but as a flexible framework that requires innovation and exposure to new information and as a consequence is different every time it is taught. This has allowed productive synergies with **indigenous and local knowledge**, and incorporation of **social technologies** neglected in early years, but now known to be vital to the success of any project.[2]

The Himalayan Permaculture Centre in Nepal (HPC) trains local farmers in permaculture design and a range of techniques of demonstrated value in improving livelihoods and supporting climate adaptation and other challenges.[3] Direct peer-to-peer learning when trainees teach other farmers what they have learnt spreads these ideas and practices far beyond the people actually present on the course. As farmers apply such techniques, they make improvements and adjustments from which HPC staff can learn, adjusting the content of taught material accordingly.[4] The Chikukwa Project in Zimbabwe partly formalises peer learning processes through organised learning visits to farms, making the sites and the farmers' knowledge and skills available as educational resources.[5] Links with international permaculture teachers, writers and students communicate knowledge and practice of local practitioners to the global permaculture community, allowing useful innovations for climate change adaptation to spread worldwide.

Permaculture courses, workshops and events are more than just sites of teaching and learning in the conventional sense. They are vital nodes in creation of communities of practice at all scales and levels, that subsequently allow exchange of new information and ideas. The Transition approach of community-led responses to climate change originated as a group design project on a permaculture course.[6] Much of its rapid spread was initially due to permaculture teachers taking it up and applying it in their own communities. As its influence spread more widely, Transition brought permaculture to vast new audiences. Its innovations in areas such as **personal resilience** and **energy descent** have changed how many people think about, teach and apply permaculture.

Gaia University delivers permaculture-based programmes in eco-social design to people all over the world using distance and virtual learning methods.[7] It promotes a culture of action-based un/learning, rooted in a willingness to document, learn from and share experiences arising from direct involvement in practical measures to address climate change and other environmental and social issues. Learners at Gaia University take on active roles in peer support, later contributing to content delivery and mentoring of others at earlier stages in their un/learning journeys. They are progressively socialised into and encouraged proactively to shape a global community of engaged change activists committed to ongoing self-reflection and collective learning.

This is a microcosm of the wider permaculture community, which in turn seeks to act as a model for the global processes of learning and change necessary for humanity as a whole to address the responsibilities that come from living in the Anthropocene, a world in which the ever-present reality of climate change presents a profound challenge to accepted ideas and practices in almost all fields of human endeavour.

Collective Intelligence at work. KATY FOX/CELL

A-frame tutorial during a permaculture course at the Central Rocky Mountains Permaculture Institute, Colorado. CRMPI

Community Garden in Luxembourg.
KATY FOX/CELL

Community Garden in Luxembourg.
KATY FOX/CELL

Esther Matirekwe at Chikukwa is a keen farmer. She grows a surplus of maize and a large crop of fruit. 2014 CHIKUKWA PROJECT/TERRENCE LEAHY

5. FUTURE STEPS

Permaculture-based responses to climate change have so far achieved only a tiny fraction of their full potential. Limited policy engagement is a major barrier to further growth and scaling up to playing a significant role in global action. New avenues are emerging for the development and implementation of policy to make fuller use of permaculture, both by providing better support to practical efforts and as a tool in policy processes themselves.

Greenhouse, Occidental Arts and Ecology Center – founded in 1994, this vibrant permaculture center hosts dozens of events each year including three plant sales offering hundreds of 100% organic heirloom vegetables, plants and flowers. Occidental, California, USA, 2015 CAROLYN MONASTRA

5.1

PERMACULTURE AND CLIMATE POLICY

Many avenues exist for productive dialogue between permaculture and climate policy at all levels, with the potential to improve the effectiveness of both

Permaculture's achievements to date supporting climate change mitigation and adaptation are in some respects impressive, though also limited. For a grassroots movement with few or no financial resources, to spread across the globe and develop workable solutions in so many different places, settings and fields of endeavour is remarkable. The strategies presented here give only a narrow and superficial view of the range of working applications. The examples used to illustrate them represent only a tiny sample of thousands of initiatives worldwide that are successfully supporting decarbonisation, adaptation to climate change and other responses to environmental, social and economic challenges.[1]

So far, these accomplishments have largely taken place at the margins of the political, economic and social mainstream. While this has allowed great freedom to innovate – an important factor in permaculture's success – it has also restricted its wider impact and possibilities for effective support at policy and other levels. The need for radical socio-economic transformation in order to address climate change and other sustainability challenges is well recognised, and is becoming a welcome and increasingly important influence on policy.[2] Permaculture is a potentially vital element of this transformation, but can fulfil this potential only through greater engagement with policy.

Structures for policy engagement are already emerging within permaculture and linked movements. The Global Ecovillage Network (GEN) has had consultative status with the United Nations Economic and Social Council since 2000, and regularly contributes to briefing sessions and committees. GEN has also participated in UNFCCC COP processes since 2007 and sent delegates to the 2012 Rio+20 Earth Summit.[3] Collaboration with local authorities has been an important part of the Transition model since the outset.[4] Emerging regional and national coordination structures for Transition now often engage with government at higher levels. In 2012, Transition Network was awarded first place in the European Economic and Social Committee Civil Society Prize.[5] Permaculture organisations, Transition groups and ecovillages, among others, in 2014 came together to establish ECOLISE, an EU-wide network of community-based sustainability initiatives that includes policy dialogue among its core pillars of activity.[6]

Preparing seedlings to plant for a Force Majeure project which involves planting seedlings at varying altitudes to see which will survive our changing climactic conditions. Sagehen Creek Field Station, Truckee, California, USA, 2014 CAROLYN MONASTRA

There is fertile ground for productive engagement between permaculture and policy. Some key steps that could help cultivate this are:

1. Support existing permaculture projects and movements to grow and flourish.

Financial resources are a key limitation for many permaculture projects. While necessary, improved funding provision is not in itself a solution. Mechanisms for allocating and using funds could build upon the financial management strategies and techniques of existing permaculture projects. Funding programmes could also take into account some of the wider innovations around business and entrepreneurship covered earlier under **regenerative enterprise** (Chapter 4.10). Beyond finance, key barriers many permaculture projects face to their success, expansion or replication are often administrative, bureaucratic, or cultural. Planning law and its implementation in the UK is a key example.[7] Property regimes are another: many flourishing projects have come to an end through loss of access to land or premises under insecure conditions of tenure. Culturally, the unconventional nature of many permaculture projects can limit their credibility in the eyes of key decision-makers unfamiliar with some of their basic premises and hence unwilling to offer support. Deeper dialogue and sustained awareness-raising are necessary, particularly around the conditions for successful **commons-based governance** (Chapter 4.11), to create the physical, administrative and conceptual space necessary for new initiatives to become established.

2 Initiate and support research on the impacts, efficacy and potential wider applications of permaculture thinking and practice.

The account of permaculture and research in Chapter 2.3 highlighted the pressing need to improve the evidence base that can inform policy interventions around permaculture. Conventional highly centralised funding regimes like the EU Framework Programmes have so far proven ineffective in addressing this; whether the reforms introduced in Horizon 2020 will remedy this remains to be seen. It will in any case be necessary to develop new approaches that can support and strengthen existing informal research efforts and create synergies with established scientific procedures and expertise. The new Permaculture International Research Network[8] and research activities being planned and coordinated through ECOLISE provide suitable frameworks to take this forward.

3 More closely link permaculture with formal education.

Permaculture has begun to have a presence in school curricula in Zimbabwe and Malawi,[9] and university curricula in several European countries. The UK Permaculture Association is working to make formal accreditation available for permaculture training at all levels. Independent education and training organisations, including Gaia Education,[10] the Integral Permaculture Academy[11] and the Permaculture College of Europe,[12] are beginning to collaborate or explore collaboration with universities on programme delivery and/or validation. **Popular education**, as Chapter 4.17 describes, has been and will certainly remain the heart of permaculture's development as a movement. However, building links with formal education at all levels will greatly increase knowledge and use of permaculture, and can contribute to the educational reforms necessary to address climate change and other sustainability challenges.[13]

4 Engage with and support emerging high-level strategic initiatives.

Creating a platform for policy dialogue around permaculture and related fields is a central pillar of the work of ECOLISE. In a separate development, at the International Permaculture Convergence in Cuba in 2013, the Permaculture's Next Big Step Project was set up to facilitate a global conversation on the aspirations and ways forward for permaculture as a worldwide movement.[14] Proposals for the next phase of this project and to support a future of greater coordination and cooperation across the movement were developed for consideration during the follow-up to the Cuba meeting, in London in September 2015.

5 Introduce permaculture thinking into policy, strategy, planning and action at all levels.

As the descriptions of Strategies in Section 4 show, permaculture is not just relevant to practical details of policy in specific fields. Its use of **social technologies** and applications in **commons-based governance** show how it can also be a methodology for more effective development and implementation of policy.

The need for more flexible, inclusive and responsive policy mechanisms to address climate change, sustainability and emerging economic challenges is nowadays well known.[15] Many strategies described in the second half of Section 4 have direct potential applications in such policy processes: for example, promoting more inclusive decision-making, enabling **conflict transformation**, supporting **personal resilience** to negotiate change, and making social insights from **bioregionalism**, **economic localisation** and **energy descent** movements more widely accessible. Strategies with a more physical focus in the first half of Section 4, and the associated techniques, have many potential applications within specific technical policy fields.

Several approaches could be taken to making these tools more available to policy processes and help to bring about the necessary changes in climate governance. Improved dialogue would bring about greater awareness and understanding of permaculture among makers and deliverers of policy. Permaculture-based training within staff orientation and continuing professional development schemes would embed permaculture skills in government at all levels.

Permaculture and policy have much to learn from and offer to each other. Creating the conditions that will allow these synergies to emerge is only one part of successful responses to climate change; it is however a vital one, more urgently needed than ever before.

Woman collecting and cleaning plastic bags by the Nairobi River. After cleaning them, she will sell some of the bags to stores to reuse and keep others to crochet products such as handbags and hats. Kenya, 2012

CAROLYN MONASTRA

5.2

CLIMATE CHANGE STATEMENT AND ACTION PLAN

From the International Permaculture Convergence 2015

In September 2015, around 700 permaculture practitioners from nearly 70 countries attended the twelfth International Permaculture Convergence in London. As part of a series of collaborative planning exercises within the Next Big Step process, a working group on climate change developed a statement and action plan. The statement, reproduced in full below, was adopted by the General Assembly of the Convergence on its closing day, September 15th. A new organisation called Permaculture Climate Change Solutions will take this work forward, providing a coherent voice for permaculture in climate change debates, supporting networking and resource sharing among permaculturalists, and facilitating linkages between permaculture practitioners and others seeking to take constructive action.[1]

Permaculture Climate Change Statement

Permaculture is a system of ecological design as well as a global movement of practitioners, educators, researchers and organizers, bound by three core ethics: care for the earth, care for the people and care for the future. Permaculture integrates knowledge and practices that draw from many disciplines and links them into solutions to meet human needs while ensuring a resilient future. With little funding or institutional support, this movement has spread over the past forty years and now represents projects on every inhabited continent. The permaculture movement offers vital perspectives and tools to address catastrophic climate change.

Human-caused climate change is a crisis of systems – ecosystems and social systems - and must be addressed systemically. No single new technology or blanket solution will solve the problem. Permaculture employs systems thinking, looking at patterns, relationships and flows, linking solutions together into synergistic strategies that work with nature and fit local conditions, terrain, and cultures.

Efforts to address the climate crisis must be rooted in social, economic, and ecological justice. The barriers to solutions are political and social, not technical, and the impacts of climate change fall most heavily on frontline communities, who have done the least to cause it. Indigenous communities hold worldviews and perspectives that are vitally needed to help us come back into balance with the natural world. We must build and repair relationships across cultures and communities on a basis of respect, and the voices, leadership and needs of frontline and indigenous communities must be given prominence in all efforts to address the problem.

Permaculture ethics direct us to create abundance, share it fairly, and limit overconsumption in order to benefit the whole. Healthy, just, truly democratic communities are a potent antidote to climate change.

Both the use of fossil fuels and the mismanagement of land and resources are driving the climate crisis. We must shift from fire to flow: from burning oil, gas, coal and uranium to capturing flows of energy from sun, wind, and water in safe and renewable ways.

Soil is the key to sequestering excess carbon. By restoring the world's degraded soils, we can store carbon as soil fertility, heal degraded land, improve water cycles and quality, and produce healthy food and true abundance. Protection, restoration and regeneration of ecosystems and communities are the keys to both mitigation and adaptation.

Permaculture integrates knowledge, experience, research and practices from many disciplines to restore landscapes and communities on a large scale. These strategies include:

- *A spectrum of safe, renewable energy technologies;*
- *Scientific research and exchange of knowledge, information and innovations;*
- *Water harvesting, retention and restoration of functional water systems;*
- *Forest conservation, reforestation and sustainable forestry;*
- *Regenerative agricultural practices: organic, no-till and low-till, polycultures, small-scale intensive systems and agroecology;*
- *Planned rotational grazing, grasslands restoration, and silvopasture systems;*
- *Agroforestry, food forests and perennial systems;*
- *Bioremediation and mycoremediation;*

- *Increasing soil organic carbon using biological methods: compost, compost teas, mulch, fungi, worms and beneficial micro-organisms;*
- *Sustainably produced biochar for carbon capture and soil-building;*
- *Protection and restoration of oceanic ecosystems;*
- *Community-based economic models, incorporating strategies such as co-operatives, local currencies, gift economies, and horizontal economic networks;*
- *Relocalization of food systems and economic enterprises to serve communities;*
- *Conservation, energy efficiency, re-use, recycling and full cost accounting;*
- *A shift to healthier, climate-friendly diets;*
- *Demonstration sites, model systems, ecovillages and intentional communities;*
- *Conflict transformation, trauma counselling and personal and spiritual healing;*
- *Transition Towns and other local movements to create community resilience;*
- *And many more!*

None of these tools function alone. Each unique place on earth will require its own mosaic of techniques and practices to mitigate and adapt to climate change.

To deepen our knowledge of these approaches and refine our ability to apply and combine them, we need to fund and support unbiased, independent scientific research.

Each one of us has a unique and vital role to play in meeting this greatest of global challenges. The crisis is grave, but if together we meet it with hope and action, we have the tools we need to create a world that is healthy, balanced, vibrant, just, abundant and beautiful.

Adopted by the General Assembly of the International Permaculture Convergence, London, 2015.

Group photo from the 12th International Permaculture Convergence at Gilwell Park, Essex, England PERMACULTURE ASSOCIATION (BRITAIN)

A permaculture skills center in Sebastapol, California CAROLYN MONASTRA

ENDNOTES

1. INTRODUCTION

1 O'Hara, E. (ed.), 2013. Europe in Transition: Local Communities Leading the Way to a Low Carbon Society. Brussels: AEIDL. www.aeidl.eu/images/stories/pdf/transition-final.pdf

2. BACKGROUND

2.1 Climate Change: from Adaptation to Transformation

1 IPCC, 2013. *Climate Change 2013. The Physical Science Basis.* Contribution of Working Group I.

2 World Bank, 2012. *Turn down the heat: why a 4°C warmer world must be avoided.* Washington DC: World Bank. documents.worldbank.org/curated/en/2012/11/17097815/turn-down-heat-4°c-warmer-world-must-avoided. IPCC, 2014. *Climate Change 2014. Mitigation of Climate Change.* Contribution of Working Group III.

3 Bast, E., S. Makhijani, S. Pickard & S. Whitley, 2014. *The Fossil Fuel Bailout: G20 Subsidies for Oil, Gas and Coal Exploration.* London: Overseas Development Institute and Washington: Oil Change International.

4 IPCC, 2014. Climate Change 2014. *Impact, Adaptation and Vulnerability.* Contribution of Working Group I.

5 http://ec.europa.eu/clima/news/articles/news_2013041601_en.htm

6 http://climate-adapt.eea.europa.eu/home

7 O'Brien, K., M. Pelling, A. Patwardhan, S. Hallegatte, A. Maskrey, T. Oki, U. Oswald-Spring, T. Wilbanks, and P.Z. Yanda, 2012. Toward a sustainable and resilient future. Pp. 437-486 in: Field, C.B. et al (eds.) *Managing the Risks of Extreme Events and Disasters to Advance Climate Change Adaptation.* Cambridge: Cambridge University Press.

8 O'Brien, K. & G. Hochachka, 2010. Integral Adaptation to Climate Change. *Journal of Integral Theory and Practice* 5(1): 89-102.

9 Holmgren, D., 2002. *Permaculture: Principles and Pathways Beyond Sustainability.* Melliodora: Holmgren Design Services.

2.2 Permaculture and Climate Change Adaptation

1 Mollison, B. & D. Holmgren, 1982 (1978). *Permaculture One. A Perennial Agriculture for Human Settlements*. Tyalgum: Tagari.

2 Benyus, J., 1997. *Biomimicry: Innovation Inspired by Nature*. New York: William Morrow.

3 Rosemund, A.D., and C.B. Anderson. 2003. Engineering role models: do non-human species have the answers? *Ecological Engineering* 20: 379-387.

4 Berkes, F., J. Colding, and C. Folke, 2000. Rediscovery of traditional ecological knowledge as adaptive management. *Ecological Applications* 10(5): 1251-1262.

5 Perkins, J., 2014. Foreword. Pp. 1-6 in Eversole, F. (ed.), *Creating a Real Wealth Economy*. Birmingham, AL: The Creative Age.

6 Steffen, W., P.J. Crutzen, & J.R. McNeill, 2007. The Anthropocene: are humans now overwhelming the great forces of nature? *AMBIO: A Journal of the Human Environment* 36(8): 614-621.

7 Rockstrom, J., Steffen, W., Noone, K., Persson, A., Chapin, F.S., Lambin, E., Lenton, T.M., Scheffer, M., Folke, C., Schellnhuber, H.J., Nykvist, B., de Wit, C.A., Hughes, T., van der Leeuw, S., Rodhe, H., Sorlin, S., Snyder, P.K., Costanza, R., Svedin, U., Falkenmark, M., Karlberg, L., Corell, R.W., Fabry, V.J., Hansen, J., Walker, B., Liverman, D., Richardson, K., Crutzen, P., Foley, J., 2009. Planetary Boundaries: Exploring the Safe Operating Space for Humanity. *Ecology and Society* 14.

8 Steffen, W., Å. Persson, L. Deutsch, J. Zalasiewicz, M. Williams, K. Richardson, C. Crumley, P. Crutzen, C. Folke, L. Gordon, M. Molina, V. Ramanathan, J. Rockstrom, M. Scheffer, H. J. Schellnhuber & U. Svedi, 2011. The Anthropocene: From global change to planetary stewardship. *Ambio* 40(7): 739-761.

9 Maffi, L. (ed.), 2001. *On Biocultural Diversity: linking language, knowledge and the environment*. Washington and London: Smithsonian Institution Press.

10 Balée, W.L. (ed.), 1998. *Advances in Historical Ecology*. New York: Columbia University Press.

2.3 Cultivating a Global Climate Change Action Research Community

1 Ferguson, R. S., & S. T. Lovell, 2014. Permaculture for agroecology: design, movement, practice, and worldview. A review. *Agronomy for Sustainable Development* 34(2): 251-274.

2 McDonald, A.J., P.R. Hobbs & S.J. Riha, 2006. Does the system of rice intensification outperform conventional best management? A synopsis of the empirical record. *Field Crops Research* 96(1): 31-36.

3 Scott, R., 2010. *A Critical Review of Permaculture in the United States*. Self-published Article. robscott.net/2010/. Accessed June 11th 2015. The author's experience trying to publish this paper – rejected several times by peer-reviewed journals on the advice of reviewers from the permaculture movement – reflects how permaculturists as well as academics perpetuate this divide.

4 Henfrey, T., 2014. Edge, Empowerment and Sustainability: Para-Academic Practice as Applied Permaculture Design. In *The Para-Academic Handbook: A Toolkit for making-learning-creating-acting*. London: HammerOn Press.

5 Sears, E., C. Warburton-Brown, T. Remiarz & R. S. Ferguson, 2013. *A social learning organisation evolves a research capability in order to study itself*. Poster presentation at the Tyndall Centre Radical Emissions Reduction Conference, London, UK, 10th - 11th December 2013.

6 Veteto, J. R. & J. Lockyer 2008. Environmental Anthropology Engaging Permaculture: Moving Theory and Practice Toward Sustainability. *Culture & Agriculture* 30: 47-58.

7 Lockyer, J. & J. R. Veteto (eds.), 2013 *Environmental Anthropology Engaging Ecotopia: Bioregionalism, Permaculture and Ecovillages*. New York and Oxford: Berghahn.

8 www.permaculture.org.uk/research/2-our-current-research-projects

9 P. Chapman, R. Sinfield & C. Warburton Brown (eds.), 2014. *The Permaculture Research Handbook*. Leeds: Permaculture Association. www.permaculture.org.uk/sites/default/files/page/document/smallhandsmall.pdf

10 http://permaculture-research.blogspot.co.uk

11 www.ipcuk.events/conference

12 www.permaculture.org.uk/research/4-international-research-network

13 Kindon, S., R. Pain & M. Kesby (eds.), 2007. *Participatory action research approaches and methods: Connecting people, participation and place*. London: Routledge.

14 http://base-adaptation.eu

15 Torres Carrillo, A., 2010. Generating Knowledge in Popular Education: From Participatory Research to the Systematization of Experiences. *International Journal of Action Research* 6(2-3): 196-222.

16 Locations include Biovilla (www.biovilla.org) and Aldeia das Amoreiras (https://centrodeconvergencia.wordpress.com/category/aldeia-das-amoreiras/)

17 www.transitionresearchnetwork.org

18 www.ecolise.eu

19 www.tess-transition.eu

20 www.drift.eur.nl/?p=7121

3. PERSPECTIVES, LOCAL AND GLOBAL

3.1 Indigenous Peoples, Climate Change, and Permaculture

1 There are also legitimate arguments for defining 'indigenous' in terms of language, culture, ethnicity, or relationship to place; some would use another term such as 'traditional' in the present context.

2 Anderson, E.N., 2010. *The Pursuit of Ecotopia. Lessons from indigenous and traditional societies for the human ecology of our modern world.* Santa Barbara: Praeger.

3 Berkes, F., 2003. Comment on Hunn et al. *Current Anthropology* 44(S): S94-S95.

4 Pretty, J., 2002. Agri-Culture. *Reconnecting People, Land and Nature*. London: Earthscan.

5 Maffi, L. (ed.), 2001. *On Biocultural Diversity: linking language, knowledge and the environment.* Washington and London.: Smithsonian Institution Press

6 Berkes, F. 2008. *Sacred Ecology: traditional ecological knowledge and resource management.* 2nd edition, revised. London: Routledge.

7 Grove, R.H., 1997. *Ecology, Climate and Empire.* Cambridge: White Horse Press.

8 Macchi, M., G. Oviedo, S. Gotheil, K. Cross, A. Boedhihartono, C. Wolfangel & M. Howell, 2008. *Indigenous and Traditional Peoples and Climate Change*. Gland: IUCN.

9 Krupnik, I. & D. Jolly (eds.), 2002. *The Earth is Faster Now. Indigenous observations of Arctic environmental change*. Fairbanks, Alaska: Arctic Research Consortium of the United States.

10 Salick, J., & A. Byg, 2007. *Indigenous Peoples and Climate Change*. Oxford: Tyndall Centre.

11 Kuecker, G.D., 2015. Enchanting Transition: a post colonial perspective. In Henfrey, T. & G. Maschowski (eds.) 2015. *Resilience, Community Action and Social Transformation*. Lisbon: FFCUL and Transition Research Network.

12 Henfrey, T. & J. Kenrick, 2015. Climate, Commons and Hope: the Transition movement in global perspective. In Buxton, N. & B. Hayes (eds.) *The Secure and the Dispossessed: How the military and corporations are shaping a climate-changed world*. London: Pluto Press.

13 Altieri, M.A., 1983. *Agroecology: the scientific basis of alternative agriculture*. Berkeley: University of California.

14 Cato, M.S., 2013. *The Bioregional Economy*. London: Earthscan.

15 Fox, K., 2013. Putting Permaculture Ethics to Work. Commons thinking, progress, and hope. Pp. 164-179 in Lockyer, J. & J.R. Veteto (eds.) *Environmental Anthropology Engaging Ecotopia*. Oxford: Berghahn.

16 Aistra, G.A., 2013. Weeds or Wisdom? Permaculture in the eye of the beholder on Latvian Eco-Health Farms. Pp. 113-129 in Lockyer, J. & J.R. Veteto (eds.) *Environmental Anthropology Engaging Ecotopia*. Oxford: Berghahn.

17 Berkes, F. & C. Folke, 1998. *Linking Social and Ecological Systems*. Cambridge: Cambridge University Press.

18 Leahy, T., 2013. *The Chikukwa Project*. gifteconomy.org.au/food-security-for-africa/the-chikukwa-project; www.thechikukwaproject.com

19 http://imapermaculture.org

3.2 Global Perspectives

1 Millennium Ecosystem Assessment, 2005. *Ecosystems and Human Well–Being*. Synthesis. Pp. 2-4. Washington, DC: World Resources Institute.

2 www.stockholmresilience.org/21/research/research-programmes/planetary-boundaries/

3 Steffen, W. et al, 2015. Planetary Boundaries: Guiding human development on a changing planet. *Science* **347** no. 6223.

4 www.stockholmresilience.org/21/research/research-news/1-15-2015-planetary-boundaries-2.0---new-and-improved.html

5 www.un.org/millenniumgoals

6 Deneulin, S. & L. Shahani, 2009. *An introduction to the human development and capability approach: Freedom and agency*. Ottawa. Kabeer, N., 2011. *MDGs, Social Justice and the Challenge of Intersecting Inequalities*. Centre for Development Policy and Research Policy Brief 3. London: School of Oriental and African Studies.

7 Muchhala, B., *Last-minute lack of transparency weakens sustainable development goals*. Guardian Newspaper, August 13th 2015. www.theguardian.com/global-development-professionals-network/2015/aug/13/lack-of-transparency-sustainable-development-goals-negotiations-united-nations

8 Hickel J., M. Kirk, & J. Brewer., *The pope v the UN: who will save the world first?* Guardian Newspaper, June 23rd 2015. www.theguardian.com/global-development-professionals-network/2015/jun/23/the-pope-united-nations-encyclical-sdgs

9 Jackson, T., 2009. *Prosperity Without Growth*. London: Earthscan.

4. STRATEGIES

4.1 Water Regulation and Management

1 Turral, H., J. Burke, & J. Faurès, 2011. *Climate change, water and food security*. Rome: Food and Agriculture Organization of the United Nations.

2 Holmgren, D., 2002. *Permaculture: Principles and Pathways beyond Sustainability*. Hepburn: Holmgren Design Services.

3 Yaholnitsky, I., 2015. *Holding the Rain*. www.fao.org/prods/gap/database/gap/files/589_HOLDING_THE_RAIN_IN_LESOTHO.PDF

4 Morrow, R. & R. Allsop, 2006. *An Earth User's Guide to Permaculture*. Pymble: Kangaroo Press.

5 Leahy, T., 2013. *The Chikukwa Project*. http://gifteconomy.org.au/food-security-for-africa/the-chikukwa-project/; www.thechikukwaproject.com

6 Yeomans, K. & P. Yeomans, 1993. *Water for Every Farm*. Southport, Queensland: Keyline Designs.

7 Permaculture Research Institute, 2008. The Jordan Valley Permaculture Project. In *The Role of Environmental Management and Eco-Engineering in Disaster Risk Reduction and Climate Change Adaptation*. http://proactnetwork.org/proactwebsite/media/download/CCA_DRR_reports/casestudies/em.report.case_9.pdf

8 Muller, B., 2015. *Tamera Healing Biotope 1*. www.tamera.org/basic-thoughts/healing-water

9 Holzer, S. & L. Dregger, 2012. *Desert or Paradise*. White River Junction: Chelsea Green.

4.2 Soil Protection and Restoration

1 FAO, 2015. *Climate Smart Agriculture Sourcebook*. Rome: Food and Agriculture Organisation.

2 UNEP 2013. *The Emissions Gap Report 2013*. Nairobi: United Nations Environment Programme.

3 Bates, A., 2010. *The Biochar Solution*. Gabriola Island: New Society Publishers.

4 Lehmann, J., J. Gaunt, & M. Rondon, 2006. Bio-char sequestration in terrestrial ecosystems – a review. *Mitigation and Adaptation Strategies for Global Change* 11(2): 395-419.

5 Leach, M., J., Fairhead, J. Fraser & E. Lehner, 2010. *Biocharred Pathways to Sustainability? Triple wins, livelihoods and the politics of technological promises*. STEPS working paper 41. Brighton: STEPS Centre.

6 Holzer, S., 2010 (2004). *Sepp Holzer's Permaculture*. East Meon: Permanent Publications.

7 Savory, A., Butterfield, J. and Savory, A., 1999. *Holistic management*. Washington, DC: Island Press.

8 Falk, B., 2013. *The Resilient Farm and Homestead*. White River Junction: Chelsea Green.

9 Teague, R., S. L. Dowhower, S. A. Baker, N. Haile, P. B. DeLaune, and D. M. Conover 2011. Grazing management impacts on vegetation, soil biota and soil chemical, physical and hydrological properties in tall grass prairie. *Agriculture, Ecosystems and Environment* **141**(3): 310-322.

10 Teague, R., F. Provenza, U. Kreuter, T. Steffens, & M. Barnes 2013. Multi-paddock grazing on rangelands: why the perceptual dichotomy between research results and rancher experience? *Journal of Environmental Management* **128**: 699-717.

11 Lovins, L.H., 2014. www.theguardian.com/sustainable-business/2014/aug/19/grazing-livestock-climate-change-george-monbiot-allan-savory

12 Roland, E. & G. Landua, 2013. *Regenerative Enterprise: Optimizing for Multi-Capital Abundance*. E-book Version 1.0. Pp. 26.

4.3 Revegetation

1 www.kusamala.org

4.4 Agrodiversity

1 Ofori-Sarpong, E. and F. Asante. 2004. Farmer strategies of managing agrodiversity in a variable climate in PLEC demonstration sites in southern Ghana. Pp. 25-37 in Gyasi, E.A., G. Kranjac-Berisavljevic, E.T. Blay & W. Oduro (eds.). *Managing Agrodiversity the Traditional Way: Lessons from West Africa in Sustainable Use of Biodiversity and Related Natural Resources*. Tokyo, New York, Paris: UNUP.

2 http://crmpi.org

3 www.pfaf.org

4 Crawford, M., 2010. *Creating a Forest Garden*. Totnes: Green Books. Crawford, M., 2012. *How to Grow Perennial Vegetables*. Totnes: Green Books.

5 Natarajan, M. & R.W. Willey, 1986. The effects of water stress on yield advantages of intercropping systems. *Field Crops Research* **13**: 117-131.

6 Zhu, Y., H. Fen, Y. Wang, Y. Li, J. Chen, L. Hu & C.C. Mundt, 2000. Genetic diversity and disease control in rice. *Nature* **406**: 718-722.

7 Van der Velden, N., 2011. *Mixed Vegetable Polycultures: Research Report*. Leeds: Permaculture Association.

4.5 Agroecology

1 Altieri, M., 1987. *Agroecology*. Boulder: Westview Press.

2 Altieri, M., 1983. *Agroecology*. Berkeley: Division of Biological Control, University of California, Berkeley.

3 Altieri, M. A., 2004. Linking Ecologists and Traditional Farmers in the Search for Sustainable Agriculture. *Front. Ecol. Environ.* 2(1): 35-42.

4 Holt-Giménez, E., 2002. Measuring farmers' agroecological resistance after Hurricane Mitch in Nicaragua: a case study in participatory, sustainable land management impact monitoring. *Agriculture, Ecosystems and Environment* **93**(1): 87-105.

5 Altieri, M. A., & P. Koohafkan, 2008. *Enduring farms: Climate change, smallholders and traditional farming communities*. Third World Network (TWN). Pp. 15-16.

6 www.independentsciencenews.org/un-sustainable-farming/how-millions-of-farmers-are-advancing-agriculture-for-themselves

7 www.foodforest.com.au/fact-sheets/animals/bettongs

8 Jacke, D. & E. Toensmeier, 2005. *Edible Forest Gardens*. White River Junction: Chelsea Green.

9 Remiarz, T., 2014. Ten Year Forest Garden Trial. Third Year Report. Leeds: Permaculture Association. www.permaculture.org.uk/sites/default/files/page/document/year_3_report_trials_v4_14-11-18.pdf

4.6 Creation and use of Microclimates

1 Holzer, S., 2010 (2004). *Sepp Holzer's Permaculture*. East Meon: Permanent Publications.

2 Bane, P., 2012. *The Permaculture Handbook*. Gabriola Island: New Society Publishers. Pp. 181-185. Ostentowski, J., 2015. *The Forest Garden Greenhouse*. White River Junction: Chelsea Green. www.crmpi.org

3 Ostenowski, J., 2015. *The Forest Garden Greenhouse*. White River Junction: Chelsea Green.

4 www.ecosystems-design.com/climate-batteries.html

4.7 Bioclimatic Building

1 www.strawbalefutures.org.uk/straw-bale-projects

2 www.laboa.org

3 A highly insulating material: a typical U value for a straw bale is around 0.13 – far below EU and all national permitted maxima. Jones. B., 2001. *Information Guide to Straw Bale Building*. Todmorden: Amazon Nails. Pp. 2.

4 Simon, K. et al, 2004. *Zusammenfassender Endbericht zum Vorhaben "Gemeinschaftliche Lebens-und Wirtschaftsweisen und ihre Umweltrelevanz."* Universität Kassel.

5 www.ecosystems-design.com/climate-batteries.html

6 Holmgren, D., 2005. *Sustainable Living at Melliodora: Hepburn Permaculture Gardens. A case study in cool climate permaculture.* 1985-2005. E-Book version 1.0. Hepburn: Holmgren Design Services. Pp. 27-28.

4.8 Energy Descent

1 Odum, H.T. & E.C. Odum, 2001. *A Prosperous Way Down*. Boulder: University Press of Colorado.

2 Jackson, T., 2009. *Prosperity without Growth*. London: Earthscan.

3 Roland, E. & G. Landua , 2013. *Regenerative Enterprise: Optimizing for Multi-Capital Abundance*. E-book v1.0.

4 Hopkins, R. (ed.), 2005. *Kinsale 2012: An Energy Descent Action Plan*. Available online at: transitionculture.org/wp-content/uploads/KinsaleEnergyDescentActionPlan.pdf

5 Hopkins, R., 2008. *The Transition Handbook*. Totnes: Green Books.

6 Holmgren, D., 2009. *Future Scenarios*. Totnes: Green Books.

7 Ward, F., A. Porter & M. Popham, 2011. *Transition Streets Final Report*. Totnes: Transition Town Totnes.

8 Beetham, H., 2011. *Social Impacts of Transition Together*. Report prepared on behalf of Transition Town Totnes.

9 Cretney, R., & S. Bond, 2014. 'Bouncing back' to capitalism? Grass-roots autonomous activism in shaping discourses of resilience and transformation following disaster. *Resilience* **2**(1): 18-31.

10 Henfrey, T. & Maschowski, G. (eds.), 2015. *Resilience, Community Action and Social Transformation*. Lisbon: FFCUL and Transition Research Network.

11 Hopkins, R., 2013. *The Power of Just Doing Stuff*. Cambridge: UIT/Green Books. Pp. 113-4.

12 www.nossasaopaulo.org.br/portal/irbem

4.9 Bioregionalism and Economic Localisation

1 Sale, K., 1983. Mother of All. *Mother of All: an Introduction to Bioregionalism*. Third Annual E.F. Schumacher Lecture. Great Barrington, MA: The E.F. Schumacher Society.

2 Lockyer, J. & J. Veteto, 2013. *Environmental Anthropology Engaging Ecotopia*. Oxford: Berghahn Books. Introduction, pp. 1-31.

3 Douthwaite, R., 1996. *Short Circuit: Practical New Approach to Building More Self-Reliant Communities*. Totnes: Green Books.

4 Cato, M.S., 2013. *The Bioregional Economy. Land Liberty, and the Pursuit of Happiness*. London: Earthscan.

5 Waddell, E., 1975. How the Enga Cope with Frost: Climatic Perturbations in the Central Highlands of New Guinea. *Human Ecology* **3**(4): 249-273.

6 www.transitionnetwork.org. Bailey, I., R. Hopkins & G. Wilson, 2010. Some things old, some things new: The spatial representations and politics of change of the peak oil relocalisation movement. *Geoforum* 41(4): 595-605.

7 Hopkins, R., 2010. *Localisation and resilience at the local level: the case of Transition Town Totnes* (Devon, UK). PhD thesis, Plymouth University.

8 www.reconomy.org/leadership-projects/evaluate-the-economic-potential-of-your-new-economy

4.10 Regenerative Enterprise

1 Morgan, T., 2013. *A Perfect Storm*. London: Tullett Prebon. www.tullettprebon.com/Documents/strategyinsights/TPSI_009_Perfect_Storm_009.pdf

2 Jackson, T., 2009. *Prosperity without Growth*. London: Earthscan.

3 Trucost PLC, 2013. *Natural Capital at Risk: the Top 100 Externalities of Business*. www.trucost.com/_uploads/publishedResearch/TEEB%20Final%20Report%20-%20web%20SPv2.pdf

4 www.reconomy.org/inspiring-enterprises

5 Lewis, M. & P. Conaty, 2013. *The Resilience Imperative*. Gabriola Island: New Society Publishers.

6 Roland, E. & G. Landua, 2013. *Regenerative Enterprise: Optimizing for Multi-Capital Abundance*. E-book v1.0.

7 Fullerton, J., 2015. *Regenerative Capitalism: How Universal Principles and Patterns will Shape our New Economy*. http://capitalinstitute.org/wp-content/uploads/2015/04/2015-Regenerative-Capitalism-4-20-15-final.pdf

8 www.novachocolate.com

9 Birnbaum, J. & L. Fox, 2014. *Sustainable Revolution*. Berkeley: North Atlantic Books. Pp.183-6.

4.11 Commons-Based Governance

1 Ostrom, E., 1990. *Governing the Commons: the Evolution of Institutions for Collective Action*. Cambridge University Press.

2 Berkes, F. & C. Folke, 1998. *Linking Social and Ecological Systems*. Cambridge University Press.

3 Davey, B., 2012. *Sharing for Survival. Restoring the Climate, the Commons and Society*. Cloughjordan: FEASTA.

4 Henfrey, T. & J. Kenrick, 2015. Climate, Commons and Hope: the Transition movement in global perspective. In Henfrey, T. & Maschowski, G. (eds.), 2015. *Resilience, Community Action and Social Transformation*. Lisbon: FFCUL and Transition Research Network.

5 Ostrom, E., 2010. The Challenge of Common-pool Resources. *Environment: Science and Policy for Sustainable Development*. 50(4): 8-21.

6 www.cell.lu

7 www.terra-coop.lu

8 www.biovilla.org

4.12 Social Technologies

1 MacNamara, L., 2013. *People and Permaculture*. East Meon: Permanent Publications.

2 Healey, T., 2013. The Chikukwa Project. http://gifteconomy.org.au/food-security-for-africa/the-chikukwa-project/ Pp. 6 in pdf version.

3 http://permaculture.com.au/macro-to-micro-introducing-eco-social-matrix-as-a-tool-for-integrated-design/

4 NSW Department of Urban Affairs and Planning, 1995. *Rural Settlement – Guidelines on Rural Settlement for the North Coast*.

5 www.cell.lu

6 www.sociocracy.info; http://participedia.net/en/methods/sociocracy

7 www.sk-prinzip.eu

8 http://openspaceworld.org/wp2/

4.13 Conflict Transformation

1 Buxton, N. & B. Hayes (eds.), 2015. *The Secure and the Dispossessed*. London: Pluto Press.

2 Leahy, T., 2013. The Chikukwa Project. gifteconomy.org.au/food-security-for-africa/the-chikukwa-project/ Pp. 8-10 in pdf version.

3 Birnbaum, J. & L. Fox, 2014. *Sustainable Revolution*. Berkeley: North Atlantic Books. Pp. 219.

4 Birnbaum, J. & L. Fox, 2014. *Sustainable Revolution*. Berkeley: North Atlantic Books. Pp. 223.

5 Irving, S., 2006. Permaculture & Peace in the Middle East. *Permaculture Magazine* 49: 26-29.

6 Birnbaum, J. & L. Fox, 2014. *Sustainable Revolution.* Berkeley: North Atlantic Books. Pp. 294-298.

4.14 Personal Resilience

1 Dickinson, J. L. 2009. The people paradox: self-esteem striving, immortality ideologies, and human response to climate change. *Ecology and Society* 14(1): 34. [online] URL: www.ecologyandsociety.org/vol14/iss1/art34/

2 Henfrey, T. & J. Kenrick, 2015. Climate, Commons and Hope: the Transition movement in global perspective. In Henfrey, T. & G. Maschowski (eds.) *Resilience, Community Action and Social Transformation.* Bristol: Good Works and Lisbon: FFCUL.

3 Dawborn, K., 2011. The New Frontier: Embracing the Inner Landscape. Pp. 3-15 in Dawborn, K. & C. Smith (eds.) *Permaculture Pioneers.* Hepburn: Melliodora.

4 Burnett, G., 2013. *Towards an Ecology of the Self. 'Zone zero zero' Permaculture Design Notes.* Westcliff on Sea: Spiralseed.

5 MacNamara, L., 2012. *People and Permaculture.* East Meon: Permanent Publications.

6 Baker, C., 2011. *Navigating the Coming Chaos: a Handbook for Inner Transition.* Bloomington: IUniverse.

7 Johnstone, C., 2011. *Find Your Power: a Toolkit for Resilience and Positive Change.* East Meon: Permanent Publications.

8 McIntosh, A., 2008. *Hell and High Water. Climate Change, Hope, and the Human Condition.* Edinburgh: Birlinn.

9 Macy, J. & C. Johnstone, 2012. *Active Hope. How to Face the Mess We're in Without Going Crazy.* Novato: New World Library.

10 Macy, J. & M.Y. Brown, 1998. *Coming back to Life. Practices to Reconnect our Lives, our World.* Gabriola Island: New Society Publishers.

11 Brown, P. & G. Carver, 2008. *Right Relationship: Building a Whole Earth Economy.* Oakland: Berrett-Koehler.

12 Haxeltine, A., and G. Seyfang, 2009. *Transitions for the People: theory and practice of 'Transition' and 'Resilience' in the UK's Transition movement.* Tyndall Centre for Climate Change Research Working Paper 134.

13 Maschkowski, G., N. Schäpke, J. Grabs & N. Langen, 2015. Learning from Co-Founders of Grassroots Initiatives: Personal Resilience, Transition, and Behavioral Change – a Salutogenic Approach. In Henfrey, T. & G. Maschowski (eds.) *Resilience, Community Action and Social Transformation.* Bristol: Good Works and Lisbon: FFCUL.

4.15 Changing Worldviews

1 Prins, G. & S. Rayner, 2007. *The wrong trousers: radically rethinking climate policy.* Oxford: James Martin Institute for Science and Civilization.

2 Hulme, M., 2009. *Why we disagree about climate change: understanding controversy, inaction and opportunity.* Cambridge University Press.

3 Norgaard, K. M., 2011. *Living in Denial: Climate Change, Emotions, and Everyday Life.* London: MIT Press.

4 Dawborn, K., 2011. The New Frontier: Embracing the Inner Landscape. Pp. 2-15 in K. Dawborn & C. Smith (eds.) *Permaculture Pioneers.* Hepburn: Melliodora Publishing.

5 Holmgren, D., 2002. *Permaculture: principles and pathways beyond sustainability.* Hepburn: Holmgren Design Services.

6 Leahy, T., 2013. *The Chikukwa Project.* http://gifteconomy.org.au/food-security-for-africa/the-chikukwa-project/ Pp. 15-16 in pdf version.

7 Birnbaum, J. & L. Fox, 2014. *Sustainable Revolution.* Berkeley: North Atlantic Books. Pp. 136-7.

8 Mehrel, L.T., 2006. Green Shalom: the new Kibbutz movement. *Dartmouth Green Magazine.* http://media.wix.com/ugd/07f815_65a69fe5a26d4ab6aba7ae538a0d8d13.pdf

9 Birnbaum, J. & L. Fox, 2014. *Sustainable Revolution.* Berkeley: North Atlantic Books. Pp. 306, 312-3.

10 K. Dawborn & C. Smith (eds.) *Permaculture Pioneers.* Hepburn: Melliodora Publishing.

11 Simos, M., 2004. *The Earth Path: grounding your spirit in the rhythms of nature.* New York: HarperCollins.

12 http://starhawk.org/permaculture-solutions-for-climate-change/

4.16 Indigenous and Local Knowledge

1 http://permacultureglobal.org/projects/84-suvraga-aguyt-co-operative

2 Birnbaum, J. & L. Fox, 2014. *Sustainable Revolution*. Berkeley: North Atlantic Books. Pp. 132-139.

4.17 Popular Education

1 http://dynamicgroups.com.au

2 Macnamara, L., 2012. *People and Permaculture*. East Meon: Permanent Publications.

3 http://permaculturenews.org/2010/01/06/farmers-handbook/

4 www.himalayanpermaculture.com

5 Leahy, T., 2013. *The Chikukwa Project*. http://gifteconomy.org.au/food-security-for-africa/the-chikukwa-project/; www.thechikukwaproject.com

6 http://transitionculture.org/essential-info/pdf-downloads/kinsale-energy-descent-action-plan-2005/

7 www.gaiauniversity.org

IMAP

5. FUTURE STEPS

5.1 Permaculture and Climate Policy

1 For further examples see Birnbaum, J. & L. Fox, 2014. *Sustainable Revolution*. Berkeley: North Atlantic Books.

2 WBGU (German Advisory Council on Global Change), 2011. *World in Transition: A Social Contract for Sustainability*. Flagship Report. Berlin: WBGU.

3 http://gen-europe.org/networks/united-nations

4 Rowell, A., 2010. *Communities, Councils and a Low Carbon Future: What We Can Do If Governments Won't*. Totnes: Green Books. Reeves, A., M. Lemon, & D. Cook, 2014. Jump-starting transition? Catalysing grassroots action on climate change. *Energy Efficiency* 7(1): 115-132.

5 www.eesc.europa.eu/?i=portal.en.events-and-activities-civil-society-prize-2012

6 www.ecolise.eu

7 Fairlie, S., 2009. Low Impact Development. People and Planning in a Sustainable Countryside. Second Edition. Charlbury: Jon Carpenter.

8 www.permaculture.org.uk/research/4-international-research-network

9 www.seedingschools.org/blog/94-permaculture-in-schools-in-southern-and-eastern-africa

10 www.gaiaeducation.net/index.php/en/gaia-education-partners/80-static-content/partners/79-academic-partnerships

11 www.integralpermaculture.org

12 http://permaculturecollege.eu

13 Orr., D., 2004. Earth in Mind. On Education, Environment and the Human Prospect. Washington: Island Press.

14 https://international.permaculture.org.uk

15 WBGU (2011). World in Transition – A Social Contract for Sustainability. Flagship Report, German Advisory Council on Global Change (WBGU). Berlin: WBGU. Driessen, P. P. J. et al, 2013. Societal Transformations in the Face of Climate Change. Helsinki: JPI Climate.

5.2 Climate Change Statement and Action Plan

1 www.permacultureclimatechange.org

FURTHER INFORMATION

Organisations and Networks

UK Permaculture Association
www.permaculture.org.uk

Permaculture Research Institute
www.permaculturenews.org

Permaculture's Next Big Step
www.international.permaculture.org.uk

Permaculture International Research Network
www.permaculture.org.uk/research/4-international-research-network

Transition Network
www.transitionnetwork.org

Global Ecovillage Network
www.gen.ecovillage.org

ECOLISE – European Community-Led Initiatives for a Sustainable Europe
www.ecolise.eu

Education

List of Permaculture Design Certificates and other training, UK and international
www.permaculture.org.uk/education/courses

Gaia University. Permaculture-based Distance Learning for social change activists
www.gaiauniversity.org

Gaia Education. Ecovillage Design Education and other international programmes
www.gaiaeducation.org

Integral Permaculture Academy. Online provider of advanced permaculture training
www.integralpermaculture.org

List of courses run internationally by the Permaculture Research Institute
www.permaculturenews.org/courses

Media

Permaculture Design Magazine (USA, formerly Permaculture Activist)
www.permaculturedesignmagazine.com

Permaculture in Practice Magazine (Australia)
www.pipmagazine.com.au

Permaculture Magazine (UK, International)
www.permaculture.co.uk

Specialist Publishers

Permanent Publications
www.permanentpublications.co.uk

Chelsea Green Publishing
www.chelseagreen.com

CAROLYN MONASTRA

Books on Permaculture

Introductory
Permaculture: A Beginner's Guide
Graham Burnett
Available from **www.spiralseed.co.uk**

Permaculture in a Nutshell
Patrick Whitefield

Intermediate
The Permaculture Way
Graham Bell

Permaculture Design
Aranya

The Permaculture Handbook
Peter Bane

Advanced
Earth Users Guide to Permaculture
Rosemary Morrow

The Earth Care Manual
Patrick Whitefield

Edible Forest Gardens (two volumes)
Dave Jacke & Eric Toensmeier

Social Permaculture and Transition
People and Permaculture
Looby MacNamara

The Transition Companion
Rob Hopkins

Theory and Philosophy
Principles and Pathways beyond Sustainability
David Holmgren

Permaculture Pioneers
Kerry Dawborn and Caroline Smith

Beyond the Forest Garden
Robert Hart

Permaculture in Practice
Sustainable Revolution
Juliana Birmbaum and Louis Fox

The Power of Just Doing Stuff
Rob Hopkins

Lead Author Biographies

Dr. Thomas Henfrey is Senior Researcher at the Schumacher Institute and Research Fellow in the Centre for Ecology, Evolution and Environmental Change (cE3c) at Lisbon University. He previously conducted PhD research on indigenous forest management and lectured in the Anthropology Department at Durham University. Tom co-founded an ecovillage in Southern Spain, holds a permaculture design certificate, and actively collaborates on research and learning with Transition Network and the Permaculture Association (Britain).

Dr. Gil Penha-Lopes, an Environmental Scientist by training, is an Invited Professor in the Science Faculty at Lisbon University and Senior Researcher at the Centre for Ecology, Evolution and Environmental Change (cE3c). He coordinates Lisbon University's involvement in the EU-funded FP7 BASE Project on bottom-up approaches to climate change adaptation, and the ClimAdaPT.Local project, which collaborates with 26 municipal councils across Portugal to devise local climate change adaptation strategies. A certified Transition Network Trainer, Gil contributes and promotes Permaculture and Transition (Towns) research in Portugal and Worldwide.

Community-led Transformations

Series Editors Thomas Henfrey and Gil Penha-Lopes.

The Community-led Transformations book series is a collaboration between members of the ECOLISE Network of European community-based sustainability initiatives and independent sustainability publishing houses, Permanent Publications and Good Works. It communicates intellectually rigorous thinking from the interface of research and practice, supporting local action and improved formulation and delivery of policy towards a fairer, more sustainable world.

Volume 1: *Permaculture and Climate Change Adaptation*. T. Henfrey & G. Penha-Lopes

Volume 2: *Resilience, Community Action and Societal Transformation*. T. Henfrey & G. Maschowski.

ECOVILLAGES
ECO-CITIES
TRANSITION TOWNS
PERMACULTURE
ECOLISE
META NETWORK
A shared platform for learning, action and advocacy, by and for community-led initiatives on climate change and sustainability in Europe
The time has come to work closer together to support community-led local action for significant change
www.ecolise.eu
ecolise
@ecolise

CAROLYN MONASTRA

Permaculture in a Nutshell

Permaculture Design
A step-by-step guide
e-book EDITION

Do it Yourself
12 Volt
SOLAR POWER

MARK BOYLE
WITH GANDHI

ENERGY REVOLUTION

EDIBLE
Perennial Gardening

How To Make A FOREST GARDEN

THE EARTH CARE MANUAL

THE LOG BOOK
e-book EDITION

AROUND THE WORLD IN 80 PLANTS

earth user's guide to permaculture

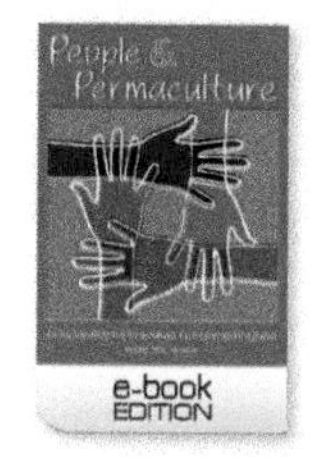
People & Permaculture
e-book EDITION

TREES for
gardens, orchards and permaculture

THE PERMACULTURE GARDEN
Graham Bell

Sepp Holzer's Permaculture

THE WOODLAND WAY
Ben Law

ROUNDWOOD TIMBER FRAMING

Subscribe to

permaculture

practical solutions for self-reliance

Permaculture magazine offers tried and tested ways of creating flexible, low cost approaches to sustainable living

Print subscribers have FREE digital and app access to over 20 years of back issues

To subscribe, check our daily updates and to sign up to our eNewsletter see:

www.permaculture.co.uk

Shipibo Youth Replanting the Amazon with commercially viable polyculture TERRA GENESIS